SpringerBriefs in Applied Sciences and Technology

SpringerBriefs present concise summaries of cutting-edge research and practical applications across a wide spectrum of fields. Featuring compact volumes of 50 to 125 pages, the series covers a range of content from professional to academic.

Typical publications can be:

- A timely report of state-of-the art methods
- An introduction to or a manual for the application of mathematical or computer techniques
- A bridge between new research results, as published in journal articles
- A snapshot of a hot or emerging topic
- An in-depth case study
- A presentation of core concepts that students must understand in order to make independent contributions

SpringerBriefs are characterized by fast, global electronic dissemination, standard publishing contracts, standardized manuscript preparation and formatting guidelines, and expedited production schedules.

On the one hand, **SpringerBriefs in Applied Sciences and Technology** are devoted to the publication of fundamentals and applications within the different classical engineering disciplines as well as in interdisciplinary fields that recently emerged between these areas. On the other hand, as the boundary separating fundamental research and applied technology is more and more dissolving, this series is particularly open to trans-disciplinary topics between fundamental science and engineering.

Indexed by EI-Compendex, SCOPUS and Springerlink.

More information about this series at http://www.springer.com/series/8884

Tiago Yuiti Kamiya ·
Marcell Mariano Corrêa Maceno · Mariana Kleina

Environmental and Financial Performance Evaluation in 3D Printing Using MFCA and LCA

Tiago Yuiti Kamiya
Production Engineering Postgraduate
Course
Federal University of Paraná
Curitiba, Brazil

Marcell Mariano Corrêa Maceno
Production Engineering Postgraduate
Course
Federal University of Paraná
Curitiba, Brazil

Mariana Kleina
Production Engineering Postgraduate
Course
Federal University of Paraná
Curitiba, Brazil

ISSN 2191-530X   ISSN 2191-5318 (electronic)
SpringerBriefs in Applied Sciences and Technology
ISBN 978-3-030-69694-8   ISBN 978-3-030-69695-5 (eBook)
https://doi.org/10.1007/978-3-030-69695-5

Jointly published with PUCPRESS - Associação Paranaense de Cultura

This Springer imprint is published by the registered company Springer Nature Switzerland AG
The registered company address is: Gewerbestrasse 11, 6330 Cham, Switzerland

Contents

Chapter 1
Printing Processes, Environmental, and Economic Evaluation

Three-dimensional printing, commonly known as additive manufacturing, allows the manufacture of a wide range of products. Among various technologies, there is a highlight for fused deposition modeling (FDM) technology, which has the largest share of its parts. Acrylonitrile butadiene styrene (ABS) and polylactic acid (PLA) are the materials mostly used in 3D printing. Other materials can be used, such as polyethylene terephthalate glycol (PETG), polyphenyl sulfone, or polycarbonate, but sophisticated technology must process these last two filaments.

The easy acquisition of 3D printers is expanding its applications, such as schools, homes, libraries, and laboratories. The advantages of this technology are related to less material waste, variety in the design of the parts, the printing of complex geometries, and the absence of molds to manufacture parts. In this manufacturing process, the extrusion of plastic filaments deposited layer-by-layer occurs. Currently, renewable materials can be used in the manufacture of filaments, and their benefits have been studied more frequently in the literature.

The use of 3D printing technology has grown significantly in recent years. According to the Kianian (2017), there was a progression in the number of manufacturers that produced and sold additive manufacturing systems, from 49 companies in 2014 to 97 companies in 2016. The additive manufacturing industry grew 17.4% in world revenue in 2016, representing $6.063 billion in the additive manufacturing industry (Kianian 2017). This growth is possible since, compared to conventional manufacturing, it can be more efficient in terms of cost and time, especially in small-scale production and customized products (Kafara et al. 2017).

However, to deal with the depletion of natural resources and environmental impacts, current manufacturing must be balanced from an ecological, social, and economic perspective (Bashkite et al. 2014). Industries usually seek to reduce the number of raw materials and time consumed in the manufacturing processes without harming finished products' performance (Ma et al. 2018).

The waste generated by additive manufacturing is potentially smaller when compared to conventional manufacturing (Rejeski et al. 2017). However, this waste

© Springer Nature Switzerland AG 2021

T. Y. Kamiya et al., *Environmental and Financial Performance Evaluation in 3D Printing Using MFCA and LCA*, SpringerBriefs in Applied Sciences and Technology, https://doi.org/10.1007/978-3-030-69695-5_1

is still present and, in some instances, much more massive than estimated due to human and machine errors (Song and Telenko 2017).

Life cycle assessment (LCA) can be used to quantify the environmental benefits of these sustainable polymers (Zhu et al. 2017). The final parts produced can also be characterized by their economic aspects (manufacture time, amount of support, and material used) (Górski et al. 2013).

The content of this material covers the environmental analysis, using the LCA method and the economic analysis using the material flow cost accounting (MFCA) for printing the same part of different materials (PLA and PETG). A method was developed for sequencing printing activities, incorporating the tools of environmental and economic analysis. Finally, the results and a comparative study between the two analyzed materials are presented.

1.1 Printing Process

Additive manufacturing is used to build a part, layer-by-layer, from a computer-aided design (CAD) model, raw material, and appropriate machines. There are numerous technologies associated with additive manufacturing, from filament extrusion to the deposition of a binder on a dust layer. These technologies differ from each other, depending on the raw material and the agglomeration process (laser, light, or liquid) (Krimi et al. 2017).

Gibson et al. (2014) define additive manufacturing as a formalized term of what was previously called rapid prototyping and which is commonly called 3D printing. Rapid prototyping is used by several companies to describe the production of a prototype before its final version or commercialization, that is, a base model for making other models before its final version (Gibson et al. 2014).

Wong and Hernandez (2012) emphasize rapid prototyping, created in 1980, as the predecessor of additive manufacturing, being the first way to create a three-dimensional object using CAD. Additive manufacturing can be divided according to its manufacturing processes, as described in Fig. 1.1. The criteria used for its classification will be according to the material in use: liquid, solid, or powder base.

The stereolithography (SL) process uses a polymerization of a photosensitive resin. In this technology, a 3D CAD model is converted into an STL format file. The platform is raised to its top, and an ultraviolet laser is used to cure the resin, transforming it into a solid layer. Then, the platform moves downward, and a new layer is built on its top, as described in Fig. 1.2 (Calignano et al. 2017).

Polyjet technology consists of building parts, layer-by-layer, by joining the inkjet technology and photopolymerization, as shown in Fig. 1.3. The first step, defined as preprocessing, determines the best orientation of the part with the manufacturing table. Then, resin drops are deposited on the table by a print head, cured layer-by-layer by ultraviolet light. The table moves down until to finish the part.

In the electron beam melting (EBM) process, a high-powered electron beam is used as an energy source instead of a laser, according to Fig. 1.4. Due to energy density

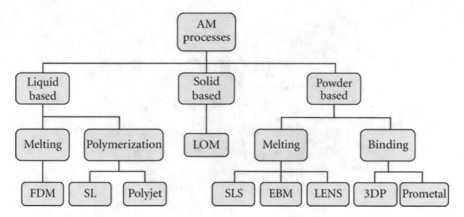

Fig. 1.1 Additive manufacturing processes. *Source* Wong and Hernandez (2012)

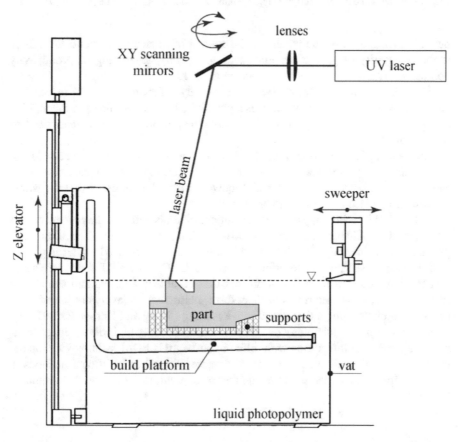

Fig. 1.2 Process and components of stereolithography technology. *Source* Calignano et al. (2017)

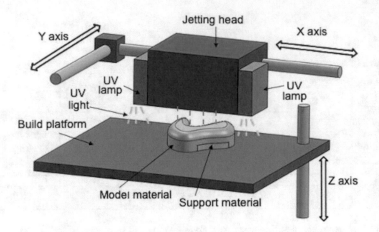

Fig. 1.3 Process and components of polyjet technology. *Source* Udroiu and Braga (2017)

being higher than laser equipment and its control from electromagnetic coils, there is the possibility of greater material melting capacity, leveraging its productivity compared to laser equipment (Calignano et al. 2017).

The selective laser sintering (SLS) technology refers to a process in which a carbon dioxide-based laser synthesizes a powder. The chamber heats to the melting temperature of the material, and the laser fuses the powder in certain areas specified in the project.

For laser engineered net shaping (LENS) technology, a part is manufactured from the injection of molten metallic powder in a predetermined location, according to Fig. 1.5. This powder fuses with a high-powered laser beam, being that the whole process takes place in a closed argon chamber.

PROMETAL technology aims to build injection tools and dies. Its process occurs when jets of a liquid binder are thrown into the stainless steel powder. The 3D printing process is a technology licensed by MIT in which a water-based binder is poured into a starch-based powder to print data from a CAD drawing. Finally, laminated object manufacturing (LOM) is a process that combines additive and subtractive techniques. The raw material is in sheet form, being that its layers are joined from pressure and heat using a thermal adhesive coating (Wong and Hernandez 2012).

Another popular technology due to its short cycle time, high-dimensional accuracy, easy use, and integration with different software is FDM technology (Boparai et al. 2016). This technology's benefits compete with other traditional methods in specific applications, thereby attracting research attention (Huang and Singamneni 2015).

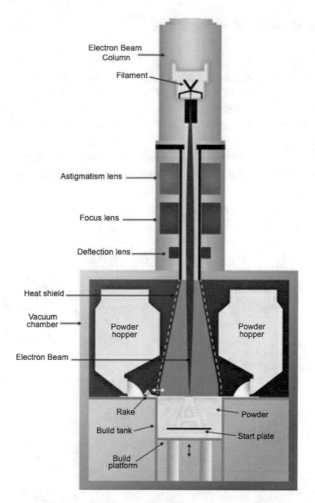

Fig. 1.4 Process and components of electron beam melting technology. *Source* Calignano et al. (2017)

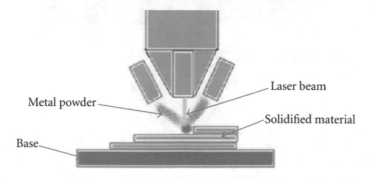

Fig. 1.5 Process and components of laser engineered net shaping technology. *Source* Wong and Hernandez (2012)

1.1.1 Fused Deposition Modeling

FDM technology uses the construction of a layer-by-layer model from the thermoplastic filament's extrusion by the movement of a nozzle (Mohamed et al. 2017). This technology created by STRATASYS (2017) uses high-performance engineering thermoplastics such as polycarbonate (PC), ABS, PC-ABS blend, polyphenylsulfone (PPSF), and nylon-12 in the construction of functional prototypes in three dimensions (Mohamed et al. 2017; Gibson et al. 2014; Tsouknidas 2011).

Figure 1.6 illustrates the operational process of FDM technology. There are two spools of filaments (build material spool and support material spool); the first is

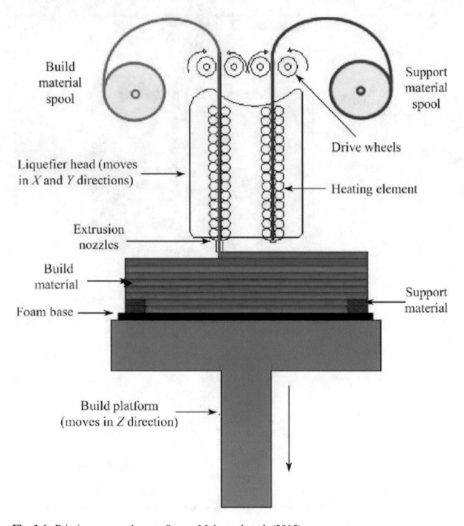

Fig. 1.6 Printing process layout. *Source* Mohamed et al. (2015)

responsible for storing the central filament. The second is for the material to assist in keeping the part during printing. Inside the head, heating elements transform the solid material into semiliquid, later expelled by extrusion nozzles. In this way, the part is built on a foam base, together with the support, with the help of the platform that moves on the z-axis (build platform) and determines the height of the built part (Mohamed et al. 2015).

Gibson et al. (2014) define the printing process by FDM technology by activities:

(a)   Activity 1: Conceptualization and CAD;
(b)   Activity 2: Conversion to STL/AMF (additive manufacturing file);
(c)   Activity 3: Transfer to additive manufacturing machine, file manipulation;
(d)   Activity 4: Configure the machine;
(e)   Activity 5: Construction;
(f)   Activity 6: Removal and cleaning;
(g)   Activity 7: Post-processing;
(h)   Activity 8: Application.

The prototyping area uses mainly FDM technology. Due to gradual improvements in materials and processes, its range of applications has increased, especially for direct use as finished parts. When producing parts for direct use, many characteristics must fulfill their functional objectives (Huang and Singamneni 2015). Mohamed et al. (2015) did a review in which the studies were carried out to determine the parameters that affect the desired characteristics of the parts. The mechanical properties, dimensional accuracy, material behavior, and surface roughness are examples of characteristics of interest and improvement in using this technology. Mohamed et al. (2017) carry out a study highlighting the significance of producing parts focusing on the contact surface, directly affecting performance in engineering applications. This technology has a wide variety of materials available and can be classified into standard materials and materials for specific applications. Among the materials considered the standard, PLA and ABS are mentioned (Boparai et al. 2016).

1.1.2 3D Printing Parameters

The requirements of high-quality parts, high productivity rate, and low manufacturing cost are essential to meet customer needs and satisfaction. The process conditions for FDM technology need to be established for each application to achieve these requirements. The selection of process parameters shows an essential role in ensuring product quality and dimensional accuracy, as well as to avoid materials waste and to reduce production costs (Mohamed et al. 2015).

Process parameters are control characteristics necessary to define the execution of activities in a process. Anitha et al. (2001) report that the value of the quality of a prototype is due to several parameters and emphasize the attempt to carry out, in the past, systematic analyses of prototypes' errors and quality. Sahu et al. (2013) emphasize the relevance of knowing the process parameters that impact a given characteristic's response. These parameters are commonly defined based on the users' experience or refer to a machine manual (default setup) (Sahu et al. 2013).

When considering the desirable characteristics of products to be printed, polymeric materials' physical properties make the FDM process complex, even considering the advancement of new materials for this technology. It highlights the value of a suitable configuration of the parameters to achieve a product's desired characteristics and the constant improvement to be made in the process during its manufacturing process (Mohamed et al. 2017).

Kuo et al. (2017) describe the importance that quality plays in the final products in mass production and many of the parameters that impact the FDM process such as

(a)   Layer thickness;
(b)   Nozzle diameter;
(c)   Envelope temperature;
(d)   Extrusion temperature;
(e)   Extrusion velocity;
(f)   Number of filling interval;
(g)   Filling velocity;
(h)   Filling pattern;
(i)   Wire-width compensation;
(j)   Feeding lag time;
(k)   Filament stopping delay time.

Mohamed et al. (2015) describe the importance of adapting the process parameters to meet the part's quality requirements. They highlight the difficulty in determining the process's ideal parameters due to many conflicting parameters that influence the properties of the materials. Figures 1.7 and 1.8 illustrate the parameters that influence the FDM process.

In Fig. 1.9, some parameters are defined when the printer follows the printing path (Mohamed et al. 2015).

Fig. 1.7 Build orientation printing a part. *Source* Mohamed et al. (2015)

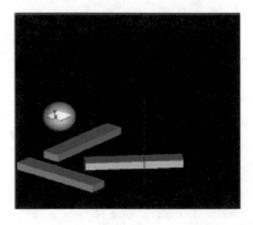

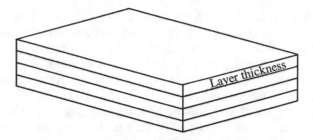

Fig. 1.8 Layer thickness in build part. *Source* Mohamed et al. (2015)

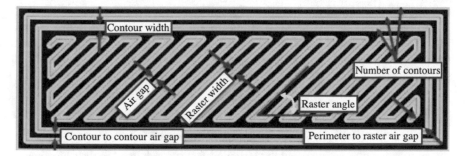

Fig. 1.9 Printing path in FDM. *Source* Adapted from Mohamed et al. (2015)

- Air gap refers to the space between one filament and another in the same layer.
- Raster angle refers to the angle between the layer and the *x*-axis. This parameter is essential in parts with small curves. It can range between 0° and 90°.
- Raster width refers to the filament width when deposited. A greater width implies a robust interior part, while a lower width will require less production time and material. This parameter ranges according to the size of the nozzle.
- Contour width refers to the width of the filament that surrounds the part.
- The number of contours refers to the number of outlines around the external and internal part's curve, and additional outlines improve the outer walls.
- Contour to contour the air gap refers to the space between the outline.
- Perimeter to raster air gap refers to the space between the innermost outline and the outlines filling edge.

1.2 Environmental Evaluation

Awareness of the sustainability of 3D printing has received increasing attention. Some points need to be addressed to assess the environmental sustainability of this technology. Three-dimensional printing processes, considered unstable, have many factors that affect their final environmental impacts, such as parameter settings and the printers themselves. The life cycle assessment (LCA) can be used to obtain an expressive conclusion quantitatively by outlining the environmental impacts of 3D printing (Liu et al. 2016). The advantage of building a highly complex part, without the additional cost involved, triggers intriguing questions for LCA, mainly of the fact that energy consumption increases due to the complexity of the material and what are the influences on the final product generated (Rejeski et al. 2017).

LCA emerged as a method capable of assessing the impacts on resource depletion, human health, and the ecosystem quality, for a product, process, or system, due to a vision that encompasses all the activities of a process, from the acquisition of raw materials to final disposal (Rejeski et al. 2017). ISO 14044: 2006 details the stages of the product life cycle. These stages are the extraction of raw materials, manufacturing, use, post-use treatment, recycling, and/or final disposal.

The phases that determine LCA studies are defined as goal and scope definition, life cycle inventory analysis (LCI), life cycle impact assessment (LCIA), and interpretation (ISO 14044 2006).

1.2.1   Goal and Scope Definition

The application, the reasons, and the target audience are defined in the objective of the LCA. The definition of the scope ensures a comprehensiveness, depth, and detail of the research to meet the outlined goal, following some items (ISO 14044 2006):

- Product system in the study;
- Functions of the product system;
- Functional unit;
- Boundaries;
- Allocation procedures;
- Impact categories and their evaluation method;
- Data requirements;
- Assumptions;
- Limitations;
- Initial requirements regarding data quality;
- Critical review, if applicable;
- Type and format of the report for the study.

1.2.2   Function, Functional Unit, and Reference Flow

The function is defined based on selecting the goal and scope, and a system can present several possibilities of functions. This function is related to the study's focus, or better, to the function/activity made by the product, process, or system in the study. The functional unit refers to quantifying the product's identified functions to provide a reference to which the inputs and outputs are related. The determination of a reference is necessary to ensure the comparison of LCA results. According to each product system studied, the reference flow determines the number of products needed to perform the function (ISO 14044 2006).

1.2.3 System Boundary

The system boundary defines the processes to be considered in the system. Based on the determination of the goal and scope, the physical system elements to be modeled are determined to obtain the level of confidence in the results and the possibility of reaching its objective. Besides, it is possible to disregard the quantification of inputs and outputs that do not significantly change the study, as long as criteria are clearly established and described (ISO 14044 2006). Once the system boundaries are defined, it is appropriate that the life cycle stages, elementary processes, and flows are determined such as

– Extraction of raw materials;
– Inputs and outputs in the main manufacturing chain;
– Production and use of fuels, electricity, and heat;
– Final disposal of waste from processes and products;
– Recovery of used products (reuse, recycling);
– Auxiliary materials manufacturing;
– Additional operations, such as lighting and heating.

1.2.4 Life Cycle Inventory

The LCI corresponds to data collection and the calculation procedure to quantify a product system's relevant inputs and outputs. This analysis has an iterative character, in which as the data collection occurs, the knowledge about the system expands. This influences the data collection procedures, and, in some instances, it helps the objective reviews or the scope of the study (ISO 14044 2006).

Data collection can be classified according to energy inputs, raw material, auxiliary inputs, products, waste, air emissions, releases to water and soil, and other environmental aspects. The calculation procedure involves validating the collected data, correlation with the elementary processes, and the correlation of the data to the reference flows and the functional unit, to elaborate the defined system's inventory results (ISO 14044 2006).

1.2.5 Life Cycle Impact Assessment (LCIA)

The LCIA studies the significance of potential environmental impacts from the results of the LCI. This phase associates the inventory data with the impact categories and specific indicators, providing data to interpret the life cycle. The LCIA has the limitation of not focusing on a complete assessment of all environmental issues in the product system under study, but only those environmental issues that have been defined in the goal and scope. The components of the LCIA phase are shown in Fig. 1.10.

The LCIA results have uncertainties related to the lack of spatial and temporal dimensions in the results of the LCI (ISO 14044 2006). Furthermore, there are no

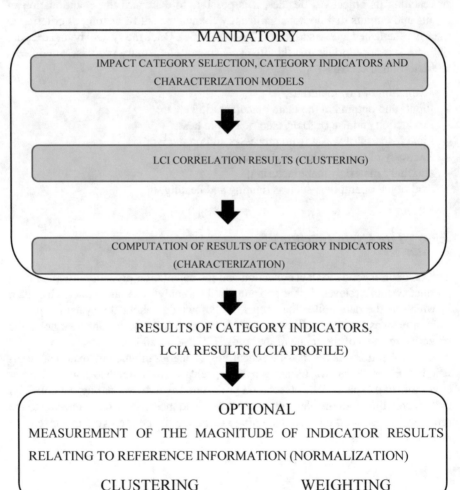

Fig. 1.10 Elements of the LCIA phase. *Source* Based on ISO 14044 (2006)

widely accepted methods that can accurately correlate inventory data with specific environmental impacts due to the impact categories being at different stages of development.

1.2.6 Interpretation of the Results Obtained in LCA Studies

The LCA interpretation is the stage at which the findings of the LCI and LCIA are carried out together to provide results consistent with the goal and enable conclusions and recommendations to decision-makers. The interpretation of the results of LCIA is based on a relative approach, which points to potential environmental effects and does not predict real impacts on category endpoints, extrapolation of limits, safety margins, or risks (ISO 14044 2006).

Nevertheless, this interpretation must support the decision toward improvements in the product's environmental performance, process, or service in the study. Considering LCA in 3D printing, its interpretation must support the decision in different aspects, such as choose among several materials/filaments to use in the 3D printing process; choose among different types of printers; among others.

1.2.7 Case Studies in LCA and Additive Manufacturing

In this section, some studies have been surveyed to show you already realized LCA's uses in additive manufacturing to support decision-makers.

In the study of Kellens et al. (2017), they evaluated LCI data and compared the impacts caused by various existing additive manufacturing technologies, including fusion and deposition modeling.

Baumers et al. (2011) carried out the same product's production using additive manufacturing technologies. They analyze the energy consumption of the product for each technology. The research aimed to demonstrate consistent and reliable results on energy consumption in the additive manufacturing process. Due to differences in the building material, layer thickness, mechanical properties, and surface finish, the information obtained was not useful for dealing with direct comparisons between the tested technologies.

Mognol et al. (2006) investigated the demand for energy consumed between three different technologies: Thermojet, FDM, and ECOS. In the rapid prototyping process, three levels were defined to classify the influence of each process parameter. The part's production time was an essential factor contributing to energy consumption, with the consumption being practically constant during the printing time. By defining acceptable process parameters, the savings in electricity consumption can reach 61% for FDM technology.

Luo et al. (1999) demonstrated a method to determine the environmental performance of freely manufactured solids. Each part of the process was divided into a life phase and analyzed separately. In the end, the effects were combined to produce the performance of the process. The characteristics considered for such an analysis thought about the respective technologies, materials, energy consumption, process waste, and destination. This comparison was made among three different additive manufacturing technologies: SL, SLS, and FDM.

Song and Telenko (2017) carried out a study to analyze the amount of material lost when a printing error occurs. The LCI was used to combine data on materials, waste, and energy consumption. It was found that the current energy consumption was 50% higher than under ideal conditions.

In summary, initial research projects on LCA have focused mainly on the subject of energy consumption. Recent research encompasses not only energy consumption as a determining factor for an LCA, but also criteria as material consumption and other categories of environmental impact. Different machines have been addressed in these new research projects (Rejeski et al. 2017).

1.2.8   Financial Evaluation

ISO 14051: 2013 defines the MFCA as a management tool that helps organizations understand the environmental and financial impacts due to material and energy and aims to seek both environmental and financial improvements for these practices. The increased transparency generated by such a tool identifies the flow of materials and energy use and the costs involved in them and compares the costs associated with products and the related costs of material loss through waste (ISO 14051 2013).

The MFCA aims to identify possible monetary gain improvements, thus avoiding unnecessary waste, residual substances, and emissions, in the summary of all non-productive materials and energy (ISO 14051 2013).

Implementing the MFCA can contribute to the knowledge of environmental and financial impacts, which increases the quality of the assessment, helping in decision-making in a process (ISO 14051 2013), such as those of 3D printing. Figure 1.11 shows the steps for implementing the MFCA.

In the planning stage (PLAN), the involvement of management is related to leading the implementation, designating those responsible, monitoring the progress, and reviewing the results of implementing an MFCA. The determination of expertise is connected to the various professionals who have information that contributes to the analysis. It includes operational expertise, engineering, quality control, environmental, and accounting. The boundary and time specifications are defined as the area to be analyzed and may cover a single process, installation, or a supply chain. The cost center's (CC) determination is related to the specified boundary steps, based on process information (ISO 14051 2013).

After the planning phase, in the application stage (DO), each cost center's inputs and outputs are identified. Materials and energy are determined as inputs, product, loss of material, or energy loss as output. From the definition of each cost center's inputs and outputs, material flows are quantified in physical units, such as mass,

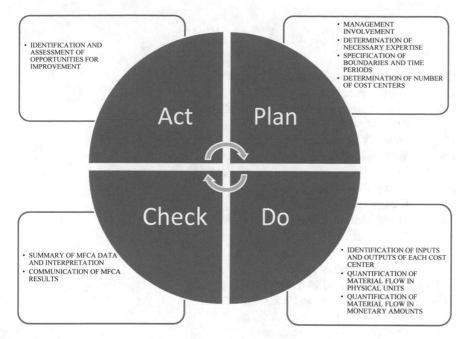

Fig. 1.11 MFCA phases for implementing the tool. *Source* Based on ISO 14051 (2013)

length, or volume, depending on the material type. When quantifying in monetary units, material, energy, system, and waste management costs are defined (ISO 14051 2013).

In this application phase, based on the data collected, it is possible to determine indicators of added value to the product or losses in waste, according to the equations mentioned in Schmidt (2014).

$$VP_i = CMp_i + CEp_i + CSp_i \tag{1.1}$$

$$mp_j \cdot VP_i = \sum_j P_{ij} \cdot QP_{ij} \cdot mp_j + mp_i \cdot pe \cdot E_i \cdot to_i + mp_i \cdot S_i \cdot ta_i \cdot n_i \tag{1.2}$$

$$VP = \sum_i VP_i \tag{1.3}$$

$$VR_i = CMr_i + CEr_i + CSr_i + CDr_i \tag{1.4}$$

$$mr_j \cdot VR_i = \sum_k P_{ik} \cdot QR_{ik} \cdot mr_j + mr_i \cdot pe \cdot E_i \cdot to_i + mr_i \cdot S_i \cdot ta_i \cdot n_i$$
$$+ mr_j \cdot DR_{ik} \cdot QR_{ik} \tag{1.5}$$

$$VP = \sum_i VR_i \qquad (1.6)$$

CMp_i	Cost of material in the product at CC_i in R\$;
CEp_i	Cost of energy in the product at CC_i in R\$;
CSp_i	Cost of system in the product at CC_i in R\$;
P_{ij}	Price of product j at CC_i in R\$;
QP_{ij}	Quantity of product j at CC_i in kg;
mp_j	Total mass of input material at CC_i;
mp_i	Material mass of output product j at CC_i;
pe	Unit price of energy per kW h;
E_i	Energy spent at CC_i in kW;
to_i	Equipment operation time at CC_i in h;
S_i	Labor cost in R\$/h at CC_i;
ta_i	Activity time per operator in h at CC_i;
n_i	Number of operators at CC_i;
CMr_i	Cost of material in waste at CC_i;
CEr_i	Cost of energy in waste at CC_i;
CSr_i	Cost of system in waste at CC_i;
CDr_i	Cost of disposal in waste at CC_i;
P_{ik}	Price of waste k at CC_i in R\$;
QR_{ik}	Quantity of waste k at CC_i in kg;
mr_j	Total mass of input waste at CC_i;
mr_i	Material mass of output waste j at CC_i;
DR_{ik}	Cost of disposal of waste k at CC_i in R\$/kg;
QR_{ik}	Quantity of waste k at CC_i in kg.

Figure 1.12 illustrates a cost center with the respective material inputs and outputs. The cost of purchased material is R\$1000, energy costs R\$50, system costs R\$800, and waste management costs R\$80 (Brazilian currency).

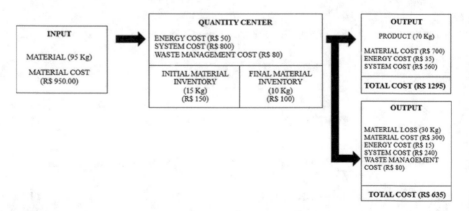

Fig. 1.12 MFCA cost center example. *Source* Based on ISO 14051 (2013)

   From this example of the quantity of input material (100 kg), 70 kg of material goes to the product, and 30 kg is transformed into a material loss. In this way, the percentage of material distribution (70% for the product and 30% of the material loss) is used to allocate energy and system costs. Still, the organization in question determines these allocation criteria. Regarding the costs of waste management, they are attributed solely to material loss, as they do not come from another nature (ISO 14051 2013).

   In the verification stage (CHECK), data obtained are summarized and exposed as a material flow cost matrix or a material flow cost diagram and subsequently communicated to stakeholders. Finally, in the assessment stage (ACTION), the results found help the organization better understand the status of the use and loss of materials, thus reviewing the data and seeing possibilities for improvement to environmental and financial performance (ISO 14051 2013).

References

R. Anitha, S. Arunachalam, P. RAdhakrishnan, Critical parameters influencing the quality of prototypes in fused deposition modelling. J Mater Process Technol **118**(1), 385–388 (2001)

V. Bashkite, T. Karaulova, O. Starodubtseva, Framework for innovation-oriented product end-of-life strategies development. Procedia Eng. **69**, 526–535 (2014)

M. Baumers et al., Energy inputs to additive manufacturing: does capacity utilization matter. Eos **1000**(270), 30–40 (2011)

K.S. Boparai, R. Singh, H. Singh, Development of rapid tooling using fused deposition modeling: a review. Rapid Prototyping J. **22**(2), 281–299 (2016)

F. Calignano et al., Overview on additive manufacturing technologies. Proc. IEEE **105**(4), 593–612 (2017)

I. Gibson, D. Rosen, B. Stucker, *Additive Manufacturing Technologies: 3D Printing, Rapid Prototyping, and Direct Digital Manufacturing.* (Springer, Berlin, 2014)

F. Górski, W. Kuczko, R. Wichniarek, Influence of process parameters on dimensional accuracy of parts manufactured using fused deposition modelling technology. Adv. Sci. Technol. Res. J. **7**(19), 27–35 (2013)

B. Huang, S. Singamneni, Raster angle mechanics in fused deposition modelling. J. Compos. Mater. **49**(3), 363–383 (2015)

ISO 14044, *Environmental Management—Life Cycle Assessment—Requirements and Guidelines* (International Organization of Standardization, 2006)

ISO 14051, *Environmental Management—Material Flow Cost Accounting—General Framework* (International Standard 14051, Geneve, 2013)

M. Kafara et al., Comparative life cycle assessment of conventional and additive manufacturing in mold core making for CFRP production. Procedia Manuf. **8**, 223–230 (2017)

K. Kellens et al., Environmental impact of additive manufacturing processes: does AM contribute to a more sustainable way of part manufacturing? Procedia CIRP **61**, 582–587 (2017)

B. Kianian, *Wohlers Report 2017*: 3D Printing and Additive Manufacturing State of the Industry, Annual Worldwide Progress Report: Chapters titles: The Middle East, and Other Countries (2017)

I. Krimi, Z. Lafhaj, L. Ducoulombier, Prospective study on the integration of additive manufacturing to building industry-Case of a French construction company. Additive Manuf. **16**, 107–114 (2017)

C. Kuo et al., A surface quality improvement apparatus for ABS parts fabricated by additive manufacturing. Int. J. Adv. Manuf. Technol. **89**(1–4), 635–642 (2017)

Z. Liu, et al., Sustainability of 3D printing: a critical review and recommendations, in *ASME 2016 11th International Manufacturing Science and Engineering Conference* (American Society of Mechanical Engineers, 2016), pp. V002T05A004–V002T05A004

Y. Luo, et al., Environmental performance analysis of solid freedom fabrication processes, in *Proceedings of the 1999 IEEE International Symposium on Electronics and the Environment, 1999. ISEE-1999* (IEEE, 1999), pp. 1–6

J. Ma et al., An exploratory investigation of additively manufactured product life cycle sustainability assessment. J. Clean. Prod. **192**, 55–70 (2018)

P. Mognol, D. Lepicart, N. Perry, Rapid prototyping: energy and environment in the spotlight. Rapid Prototyping J. **12**(1), 26–34 (2006)

O.A. Mohamed, S.H. Masood, J.L. Bhowmik, Optimization of fused deposition modeling process parameters: a review of current research and future prospects. Adv. Manuf. **3**(1), 42–53 (2015)

O.A. Mohamed, S.H. Masood, J.L. Bhowmik, A parametric investigation of the friction performance of PC-ABS parts processed by FDM additive manufacturing process. Polym. Adv. Technol. **28**(12), 1911–1918 (2017)

D. Rejeski, F. Zhao, Y. Huang, Research needs and recommendations on environmental implications of additive manufacturing. Additive Manuf. **19**, 21–28 (2017)

R.K. Sahu, S.S. Mahapatra, A.K. Sood, A study on dimensional accuracy of fused deposition modeling (FDM) processed parts using fuzzy logic. J. Manuf. Sci. Prod. **13**(3), 183–197 (2013)

M. Schmidt, The interpretation and extension of material flow cost accounting (MFCA) in the context of environmental material flow analysis. J. Clean. Prod. **108**, 1310–1319 (2014)

R. Song, C. Telenko, Material and energy loss due to human and machine error in commercial FDM printers. J. Clean. Prod. **148**, 895–904 (2017)

STRATASYS, *Tecnologia FDM*. Available: https://www.stratasys.com/br/impressoras-3d/techno logies/fdm-technology. Accessed in: 25 May 2017

A. Tsouknidas, Friction induced wear of rapid prototyping generated materials: a review. Adv. Tribol. **2011**, 1–7 (2011)

R. Udroiu, I.C. Braga, Polyjet technology applications for rapid tooling, in *MATEC Web of Conferences* (EDP Sciences, 2017), pp. 03011

K.V. Wong, A. Hernandez, A review of additive manufacturing. ISRN Mech. Eng. 1–10 (2012)

Z. Zhu, N. Anwer, L. Mathieu, Deviation modeling and shape transformation in design for additive manufacturing. Procedia CIRP **60**, 211–216 (2017)

Chapter 2
Methodology of Environmental and Financial Performance Evaluation in 3D Printing

The evaluation of financial and environmental performance is important when it comes to supporting decision-making in production processes, regardless of the nature of the process, that is, regardless of whether the process is related to an industrial, commercial, service activity, among others.

When we talk about 3D printing processes, they can have different applicability, such as the production of components of a product, the production of marketable products, or even the production of test products. Besides, they can have different application scales, such as industrial or even residential. And they can also present different technologies used, such as fused deposition modeling (FDM), stereolithography (SL), selective laser sintering (SLS), among others already mentioned in Chap. 1.

The life cycle assessment (LCA) is used as a popular and standardized method that allows assessing the environmental aspect of products and processes (Ingrao et al. 2016), in addition to providing a structure to quantify the potential environmental impacts throughout its life cycle (Lam et al. 2020).

That is, combined with the expansion of 3D printers in the industrial sphere and the concern about economic and environmental aspects, the evaluation of financial and environmental performance becomes a relevant issue.

In this sense, this chapter deals with the presentation of the methodology of environmental and financial performance evaluation in 3D printing processes. The steps of the methodology are shown in Fig. 2.1.

2.1 Identification of the Problem and Definition of Printing Parameters

The first step in the methodology reflects the identification of the problem. The problem is related to the focus of analysis. Considering the variations in 3D printing

© Springer Nature Switzerland AG 2021

T. Y. Kamiya et al., *Environmental and Financial Performance Evaluation in 3D Printing Using MFCA and LCA*, SpringerBriefs in Applied Sciences and Technology, https://doi.org/10.1007/978-3-030-69695-5_2

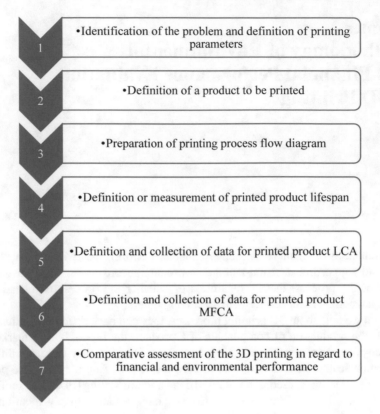

Fig. 2.1 Sequence of steps of the methodology for environmental and financial performance evaluation in 3D printing. *Source* The authors (2020)

technology and the possible variations in materials and setup for printing, they can result in the problem to be analyzed concerning the comparison of financial and environmental performance such as

- Comparison of different 3D printing technologies;
- Comparison of different printing materials for the same 3D printing technology;
- Comparison of other print setup parameters for the same material and same 3D printing technology.

In addition to defining the problem, another critical stage in this step of the methodology is the setup of printing parameters, that is, what are the parameters in the 3D printing process that affect the dimensional accuracy of a specific printed part.

Table 2.1 gives an idea of possible printing parameters for FDM technology, considering surveys in the literature.

To define the parameters in the printing process, the following actions can be taken:

Table 2.1 Printing parameters for FDM in the literature

Authors	Influent parameters
Kumar and Regala (2012)	Layer thickness, orientation, raster angle, raster width, and air gap
Dixit et al. (2016)	Nozzle diameter, slice height, and raster width
Mohamed et al. (2016)	Layer thickness, air gap, raster angle, build orientation, and the number of contours
Zhang and Peng (2012)	Wire-width compensation, extrusion velocity, filling velocity, and layer thickness

Source The authors (2020)

(a) Default machine definition, considering that the printing parameter is not part of the defined problem;

(b) Definition of parameters based on the literature review about studies that have already done dimensional accuracy tests for the printing technology chosen in the problem; or

(c) Parameter definition based on practical printing tests and dimensional accuracy measurement.

In this sense, it is up to the technology's user to opt for one of the three methods mentioned above.

After this first step, the product to be printed must be defined. This definition is detailed in Sect. 2.2.

2.2 Definition of a Product to Be Printed

The definition of a product for printing is essential for the evaluation of financial and environmental performance since it is from this product that the data to be used by the material flow cost accounting (MFCA) and life cycle assessment (LCA) tools are obtained.

The variations of products for printing are diverse, and many of the possibilities are known to the user in the 3D printing process, that is, printing products can be

- finished and marketable products from 3D printing;
- components that are part of a finished and marketable product;
- products used as a test for finished products; or
- auxiliary products in the production processes.

If the user does not have a product already used in the printing routine, it is recommended that he defines a product that has different dimensional elements, such as curves, hollow areas, among other characteristics. This action must be reasoned because a more complex part to be printed to highlight differences in performance, as it is amid materials, printing technologies, or even printing parameters, facilitating the comparison of financial and environmental performance.

2.3   Preparation of Printing Process Flow Diagram

The printing process flow diagram must be prepared based on the defined problem, which contains the type of technology and the printing materials to be used. In this context, the stages of the printing process, such as product design in computer-aided design (CAD) software, a supply of printing material, table heating, among other steps, must be considered.

Also, one can consider the indication of steps before and after the 3D printing process from a product life cycle perspective.

From this stage of the methodology, it is recommended to obtain a process flow diagram design, with detailed identification of the stages.

2.4   Definition of the Printed Product Life Span

When the product to be printed is defined, the product's life span must be obtained to estimate its life cycles and consequently provide a comparative basis for the defined study problem.

This life span can be obtained in two ways as follows:

- By monitoring the wear of the printed product by recording the frequency of exchanges or
- Through wear test of the printed product.

Whenever it is possible, it is recommended to use the product's wear time obtained from exchange frequency records since this time represents the reality of the printed product's durability.

Despite that, in cases where the user of the 3D printing technique does not have this information, an alternative involves the product wear test. This type of test is characterized by the exposure of the product to continuous and defined cycles of use, with the measurement between cycles of wear of this product. Through this measurement, it is possible to draw wear trend curves and consequently estimate the life span of the printed product.

The life span of the part was defined based on functional tests simulating the wear of the gap gages. These tests were carried out from the process of using a gage sequentially, checking in which regions there was greater wear of this part until it was considered not useful because it was out of technical specification. Thus, it was found that the region of greatest wear corresponded to the region with thickness limits between 2.45 and 2.75 mm.

2.5 Definition and Collection of Data for Printed Product LCA

ISO 14044: 2006 details the integral stages of a product's life cycle, namely the extraction of raw materials, production, use, post-use treatment, recycling, and final disposal.

The stages that determine LCA studies are the definition of objective and scope, life cycle inventory analysis (LCI), life cycle impact assessment (LCIA), and interpretation (Fig. 2.2).

2.5.1 Definition of Objective and Scope

The objective of the LCA must contain the intended application, the reasons for its realization, and the target audience for which the study is intended. The interest in using the results in comparative statements for public disclosure must also be indicated.

The intended application corresponds to the application focus of the LCA study, that is, what is the application interest? This interest will be related to the definition of the problem presented in Sect. 2.1 of this chapter, and it will often involve the

Fig. 2.2 LCA stages. *Source* Based on ISO 14044 (2006)

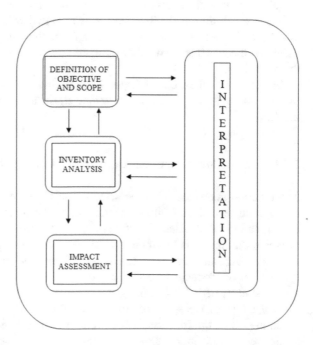

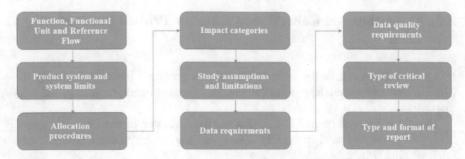

Fig. 2.3 Scope definition stages in LCA. *Source* Based on ISO 14040 (2006)

comparison, either of printing technologies, materials used in printing, or even of printing parameters.

The reasons for the LCA study correspond to the motivation for this study. The reasons appear as a complement to the intended application. It will also be related to the definition of the problem in Sect. 2.1 and will be related to the answer to be sought in the LCA as follows: Which technology has the least potential environmental impact? Which printing material has the least potential environmental impact? What printing do parameters provide a printed product with the least potential environmental impact?

Finally, the target audience is nothing more than those interested in the study. Usually, it will be the users of 3D printing who are developing the study.

The scope indicates the 3D printing system that will be analyzed, delimiting its borders and other functional characteristics of the study (Elcock 2007). Thus, its definition ensures that the depth, extent, and width of the study are plenty to achieve the objectives.

In this context, the scope of the LCA can be divided into nine stages, as established by ISO 14040: 2006 (Fig. 2.3). These nine stages are detailed from Sects. 2.5.1.1 to 2.5.1.6.

2.5.1.1   Function, Functional Unit, and Reference Flows

The function is defined from the selection of the object and scope since a system can contain several possibilities of functions. The functional unit characterizes the quantification of the identified functions of the product to provide the reference to which the inputs and outputs are related. The determination of the reference is necessary to ensure the comparison of the LCA results. The reference flow determines the number of products needed to perform the function, according to each product system studied (ISO 14044 2006).

In the case of the types of studies possible for this methodology, the product function will be the function of the product to be printed. The functional unit will be defined or measured life span for the product, according to Sect. 2.4, and the reference flow will be the number of printed products needed to meet the functional unit.

2.5.1.2   Product System and System Limit

The product system can be defined as the "set of elementary processes, connected materially and energetically by the elementary, intermediate, and product flows, which models the life cycle of a product." (Da Silva et al. 2015, p. 44). That is, it corresponds to the mass and energy flows of the life cycle under analysis considered in the life cycle assessment of the 3D printing process under study.

This system, according to its extension, can be divided into variant approaches. These approaches involve the definitions of

(a)   Cradle: The cradle means the beginning of a life cycle, characterized by the extraction of natural resources. It can be said that it represents the birth of products or services.
(b)   Grave: The grave means the end of a life cycle assessment. It is related to the death of a product or service, representing their final destination.
(c)   Gate: The gate means any intermediate stage between the cradle and the grave within a life cycle under analysis.

For the LCA of 3D printing based on the problem defined in Sect. 2.1, it is recommended to use the cradle to grave product systems.

This system approach refers to life cycles that consider everything from the stages of extraction of raw materials for the product/service under analysis to its elimination/disposal. In other words, it represents everything from the extraction of the component materials of the 3D printing process to the disposal of the printed product.

The limit of a system represents the stages of the life cycle assessment that are not being considered in the analysis. Usually, these limits are adopted due to the lack of information available for the quantification of environmental impacts over the life cycles under study.

2.5.1.3   Allocation Procedure

The allocation procedures correspond to a proportional distribution of the inputs and outputs of a process or product system between the product system under study and other product systems. As recommended by ISO 14040 (2006), allocation procedures should be avoided. In this sense, this methodology will follow the recommendation of the standard for the evaluation of environmental performance with allocations.

2.5.1.4 Impact Categories

The impact categories can be defined as "classes that represent the relevant environmental issues to which the results of the analysis of the life cycle inventory can be associated" (ISO 14044 2006, p. 5). Typically, the impact categories are already defined when choosing a LCIA method, which corresponds to LCA stage 3. There are several methods of LCIA available in the literature, as can be seen in Carvalho et al. (2014).

This methodology recommends the use of LCIA methods that provide potential impact responses in the form of an Ecopoint (Pt) to facilitate support for the 3D printing user's decision, such as IMPACT 2002+, ReCiPe, and ILCD. The choice of method is at the discretion of the user. However, a literature background is recommended for this choice.

2.5.1.5 Study Assumptions and Limitations

The assumptions and limitations refer to considerations that are made throughout the LCA under analysis to make it feasible for measuring the potential environmental impact.

Commonly, the assumptions and limitations are related to life cycle assessment information under analysis that are disregarded by the LCA due to the lack or difficulty of obtaining reliable numerical data, and that the non-consideration of this information tends not to significantly influence the final results of the LCA. That is, these assumptions and limitations are often linked to the definition of a cut-off criterion for data. In this sense, this methodology recommends the use of a data cut-off criterion of 1% of the total mass of processes input, a value that is commonly applied, according to Passuello et al. (2014).

2.5.1.6 Data Requirements, Data Quality Requirements, Type of Critical Review, and Report Type and Format

The last four stages of the scope of an LCA are not considered in this methodology.

2.5.2 Life Cycle Inventory (LCI) Analysis

The LCI analysis corresponds to the stage of collecting data on the input and output of materials and energy along the 3D printing life cycle (LC) under study. This stage is where data collection is carried out, and the inventory of materials and energy is set up.

It is important to emphasize that this survey must contemplate the definitions already made in the objective and scope stage, such as product system, system limits, assumptions, function, functional unit, and reference flow, among others.

The data that can be used to compose an LCI vary between three different types, namely collected data, database data, and simulated/modeled/calculated data.

In this sense, for this methodology, it is recommended that the data directly related to the 3D printing process, such as the mass input and output quantities of the printing materials and the energy consumption of the equipment in the printing process, are primary, that is, data collected at the source.

To measure the life span of the printed product, it is recommended to use the data obtained in Sect. 2.4.

Finally, for other data, it is recommended to use a database, such as Ecoinvent.

2.5.3 Life Cycle Impact Assessment (LCIA) and Interpretation

The interpretation of the LCA is provided by associating the inventory data with the impact categories and specific indicators through the LCIA method.

To obtain the results of this stage, it is recommended to use the LCIA method defined in Sect. 2.5.1.4. The use of the single score in LCIA methods facilitates the decision process to meet the defined 3D printing LCA objective.

In this sense, the interpretation of the results of potential environmental impact is limited by analyzing the single score values, with the highest values representing the highest potential environmental impacts and the lowest values, consequently, the lowest potential environmental impacts.

2.6 Definition and Collection of Data for Printed Product MFCA

The identification of possible improvements in monetary gain is made by the MFCA process, thus avoiding waste of non-productive materials and energy (ISO 14051 2013). Schmidt (2014) highlights the MFCA for the reason of not only showing the direct costs of waste (losses) but the losses of values within the company, including the costs of material, labor, and capital.

MFCA can provide meaningful information at various stages of a Plan-Do-Check-Action cycle, described in Sect. 1.2.8.

2.7   Comparative Assessment of 3D Printing Financial and Environmental Performance

The comparative assessment of the economic and environmental performance is obtained using the total results found after the simulation in the SimaPro software, as well as the results of the MFCA. Through these results, a sequence of steps is designed to assess the financial and environmental performance in 3D printing processes (Fig. 2.4).

From the workflow allied to the LCA and MFCA steps, it was possible to create the sequencing of the environmental and economic performance evaluation for the 3D printing process. The step-by-step describes from the identification of the problem and definition of the parameters to be used in printing, followed by the definition of the material to be used and which product will be printed. The printing flowchart is followed based on the previous definitions to understand the inputs and outputs of each step and the controllable and non-controllable parameters existing in the printing process. The measurement of life span is related to the definition of reference flows, a specific characteristic in each study to be carried out. From these definitions, they are broken down into activities in parallel, characterized by the methods of LCA and MFCA. The results obtained from following the four stages of LCA, in convergence with the results obtained from the simulation of MFCA, make it possible to carry out a comparative analysis of these and further guidance when related to environmental, economic, or economic and environmental performance.

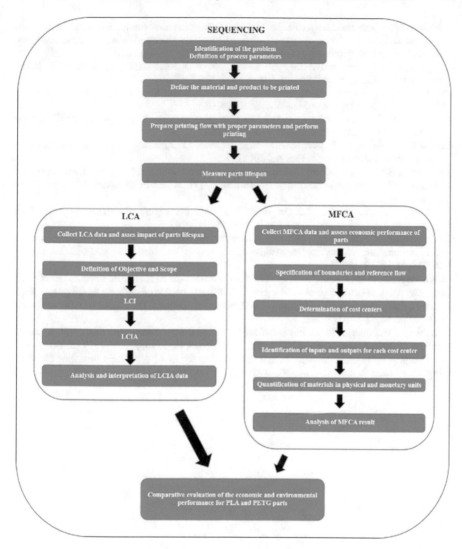

Fig. 2.4 Sequencing of the evaluation of environmental and economic performance in 3D printing processes. *Source* The authors (2020)

References

A. Carvalho et al., From a literature review to a framework for environmental process impact assessment index. J. Clean. Prod. **64**, 36–62 (2014)

G.A. Da Silva, M. Bräsher, J.A. Oliveira Lima, C.R. Lamb, *Avaliação do Ciclo de Vida—Ontologia terminológica. Brasília* (DF: IBICT, 2015)

N.K. Dixit, R. Srivastava, R. Narain, Comparison of two different rapid prototyping system based on dimensional performance using grey relational grade method. Procedia Technol. **25**, 908–915 (2016)

D. Elcock, *Life-Cycle Thinking for the Oil and Gas Exploration and Production Industry. Argonne National Laboratory.* Technical Report ANL/EVS/R-07/5 (2007). Available: https://www.evs. anl.gov/publications/pubdetail.cfm?id=60178. Accessed in: 21 June 2018

C. Ingrao et al., A comparative life cycle assessment of external wall-compositions for cleaner construction solutions in buildings. J. Clean. Prod. **124**, 283–298 (2016)

ISO 14040, *Environmental Management—Life Cycle Assessment—Principles and Framework* (International Organization of Standardization, 2006)

ISO 14044, *Environmental Management—Life Cycle Assessment—Requirements and Guidelines* (International Organization of Standardization, 2006)

ISO 14051, *Environmental Management—Material Flow Cost Accounting—General Framework* (International Standard 14051, Geneve, 2013)

G.P. Kumar, S.P. Regalla, Optimization of support material and build time in fused deposition modeling (FDM), in *Applied Mechanics and Materials* (Trans Tech Publications, 2012), pp. 2245–2251

K.L. Lam, L. Zlatanović, J.P. Van Der Hoek, Life cycle assessment of nutrient recycling from wastewater: a critical review. Water Res. **173**, 115519 (2020)

O.A. Mohamed, S.H. Masood, J.L. Bhowmik, Mathematical modeling and FDM process parameters optimization using response surface methodology based on Q-optimal design. Appl. Math. Model. **40**(23–24), 10052–10073 (2016)

A.C.B. Passuello et al., Aplicação da avaliação do ciclo de vida na análise de impactos ambientais de materiais de construção inovadores: estudo de caso da pegada de carbono de clínqueres alternativos. Ambiente Construído: Revista Da Associação Nacional De Tecnologia Do Ambiente Construído **14**(4), 7–20 (2014)

M. Schmidt, The interpretation and extension of material flow cost accounting (MFCA) in the context of environmental material flow analysis. J. Clean. Prod. **108**, 1310–1319 (2014)

J.W. Zhang, A.H. Peng, Process-parameter optimization for fused deposition modeling based on Taguchi method, in *Advanced Materials Research* (Trans Tech Publications, 2012), pp. 444–447

Chapter 3
Case Study Applying the Methodology in a 3D Printing Process

This book is structured around the assessment of financial and environmental performance in the 3D printing process. In this sense, this chapter's objective is to demonstrate the application of the methodology presented in the previous chapter through the use of a case study of printing a part with two different types of filaments. This case study evaluates comparatively financial performance through the MFCA technique and environmental through the LCA technique of these two different filaments (PLA and PETG). The chapter is organized according to the methodological sequence already presented in Chap. 2.

3.1 Characterization of the Study's 3D Printer

The case study presented exemplifies the methodology of evaluating financial and environmental performance in 3D printing processes comprised the use and evaluation of FDM technology without using a heated table, with two types of filament materials, being PLA and PETG.

The equipment and software used in this study follow Table 3.1.

It is worth mentioning that the hardware and software definitions adopted in this book exemplify the application of the methodology for assessing financial and environmental performance, using the techniques of LCA and MFCA. However, the researcher/user may choose to use other printer models and printing software for replicating this experiment.

The printer manufacturer already has standard settings that help determine the settings that best fit each material regarding the printing parameters.

In this sense, in this case study, the filament width, when deposited, was defined as the control variable, keeping all other variables constant. This parameter was set because the printer does not have the automatic leveling of its table, and adjustments are needed to determine the parts' height. Many printers already have automatic

© Springer Nature Switzerland AG 2021
T. Y. Kamiya et al., *Environmental and Financial Performance Evaluation in 3D Printing Using MFCA and LCA*, SpringerBriefs in Applied Sciences and Technology, https://doi.org/10.1007/978-3-030-69695-5_3

Table 3.1 Equipment and software used

Item	Name	Description
Printer	Stella 1	FDM printer, manufactured by the company Boa Impressão 3D
Software	Repetier-Host 2.0.5	The software of Hot-World GmbH & Co. KG used to host print files
Slicing software	Slic3r v1.6	The software of Alessandro Ranellucci used to perform the slicing of the 3D part for later implementation of the 3D printing process

Source The authors (2020)

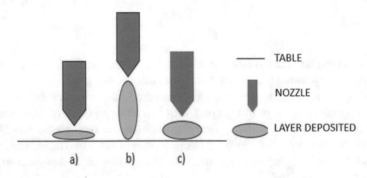

Fig. 3.1 Influence of parameter z offset. *Source* The authors (2020)

leveling attached to the printer, or the possibility of inserting this mechanism later, helping the user obtain greater precision when adjusting this parameter. In the printer used in this case study, this filament width is defined by the parameter of z. This parameter determines how much the extruder will travel on the z-axis. High values of z make the width of the filament smaller, and shallow values of z tend to flatten the layer and make the width of z larger.

In Fig. 3.1, three variations of the z value are shown, with item (a) referring to the crushing of the layer when the z offset value is low, item (b) referring to the layer when the z value is high, tending to the detachment of the layer, and the item (c) is defined as an ideal layer for printing.

Following the case study presentation, we continue with the details of the printed product and the focus of the comparative assessment.

3.2 The Printed Product

The product used in the case study was the gap gauge. It is a measuring device for controlling distances between two parts. It is used to manufacture and maintain automotive equipment, aviation, calibration devices, and services. They are blades

Fig. 3.2 Gap gauge used in the case study. *Source* The authors (2020)

of different diameters to adjust gaps to enable the correct assembly sets. They can be used in adjusting tappets, spark plugs, checking gaps in bearings and gears, adjusting pistons, rings, in addition to controlling the gap of various equipment (ÔMICROM 2018).

In the automotive industry that was part of the case study, the gap gauge is used to control the car door and body gap. In the car assembly process, the correct fixing of the doors inhibits the possibility of quality failures. The gap gauge is essential as a device to ensure the assembly of this set. Figure 3.2 shows the gap gauge used in the case study. It is worth mentioning that the partner industry in the case study was not identified in this book due to the privacy right adopted by the company.

3.3 Flowchart 3D Printing of the Gap Gauge

The 3D printing flowchart construction started by studying and understanding the functioning of 3D printing through FDM technology.

In this sense, from the analysis of the studies by Kumar and Rugala (2012), Dixit et al. (2016), Mohamed et al. (2016), and Zhang and Peng (2012), who carried out experiments focusing on dimensional characteristics, and the flowchart for the FDM printing process was prepared. The structure was defined based on an adaptation by Gibson et al. (2014), highlighting the printing process stages.

After the printing activities survey, the 3D printing flowchart of the case study was graphically constructed. The software used for the construction of the flowchart was R v3.4.3, and the code was an adaptation of the process mapping model of Cano et al.

(2012). This software is open-source, making it possible to change data according to the user's needs, but using other software that performs process mapping.

Figure 3.3 shows the flowchart adapted from Gibson et al. (2014), in which the inputs and outputs for each process are defined. The "Param (x)" determines the parameters that can be controlled at each phase's input. For example: In the model creation step, the part design and product design parameters define an output, characterized by "Featur (y)", in this example of the part design. The letters represent the type of characteristic, with (N) being defined as noise, in this case, the filament density, as they are characteristics external to the process, depending on the material to be used. The (P) is defined as a procedure found in the drawing's design, in which for the assembly of a part, there are several ways to conceive it and procedures involved in this task. (Cr) is defined as critical, not evaluated in this study. The (C) is related to the controllable parameters that have adjustments to be performed in the steps.

Finally, it is worth mentioning that the construction sequence of the printing flowchart presented here can be adapted to other additive manufacturing processes and input and output data and system noise.

3.4 Print Test, Final Print, and Part Life

The 3D printing process started with the printing test. This test was performed considering the printer configuration, adopting the standard configuration with adjustment of the z offset value, and the dimensional specifications of the part to be printed, the gap gauge. Besides, other characteristics of the printer and the filament used were the filaments used for the tests have a diameter of 1.75 mm; the printer does not have automatic leveling of its base, and the printing dimensions of its table are $200 \times 200 \times 200$ mm; the extruder nozzle used had a diameter of 0.4 mm, which allows the accuracy of 0.05 mm for the layer height and 0.01 mm for positioning accuracy.

In this sense, the gap gauges, to be considered compliant, must meet the thickness specifications defined by the automotive partner industry. Part thicknesses must be within a tolerance range, ranging from 2.45 to 2.75 for nominal thicknesses of 2.60 mm. For the thickness of 3.6 mm, the parts must meet the range of 3.45–3.75.

The first test was carried out with a z offset value of 1.2 mm, but the result on both parts was non-compliant, presenting a thickness of 2.79 mm. In the sequence, tests were carried out, adjusting the z offset value for each test by -0.1 mm, as the lower layers did not come out with the desired quality. After the tests, the z offset that presented the final part was adjusted to 0.9 mm.

With z offset defined, it was possible to start definitive part printing to control gaps between doors and car bodies by the automotive industry.

With the final parts printed, the definition of part life began, aiming to apply the LCA and MFCA tools in a product system from the cradle to the grave.

The useful life of the part was defined based on functional tests simulating the gap gauges' wear. These tests were carried out using a gauge sequentially, checking in

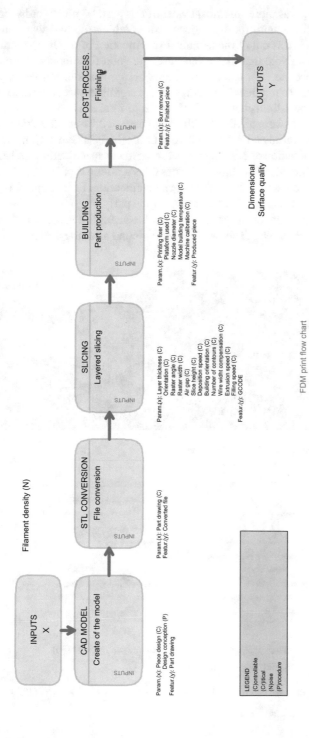

Fig. 3.3 Flowchart of inputs and outputs for each process. *Source* Adapted from Gibson et al. (2014)

which regions there was more significant wear of this part. In other words, the gauge was used several times until the wear occurred, in which the part was considered not useful because it was out of the technical specification of use by the industry to perform its function. Note that these wear tests are valid for this book's study, and the user must choose the wear test that best suits his application, if necessary, in his study. Finally, it was found that the region of most significant wear corresponded to the area with thickness limits between 2.45 and 2.75 mm, as shown in Fig. 3.4.

The definitive parts were obtained with a thickness of 2.71 and 3.72 mm in thickness for the PLA material and 2.69 and 3.70 mm for the PETG material. The gauge has a nominal specification of 2.60, 2.45 mm for the lower specification, and 2.75 mm for the upper specification. The other thickness has a nominal of 3.60, 3.45 mm for the lower specification, and 3.75 for the upper specification. From Figs. 3.5, 3.6, 3.7 and 3.8, data plots are shown after wear tests of the parts.

From the equation generated for each curve, the useful life values for the printed parts were estimated. Table 3.2 shows the results found after the tests.

Fig. 3.4 Highest wear region. *Source* The authors (2020)

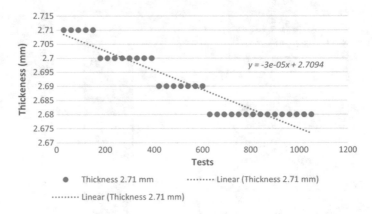

Fig. 3.5 Wear of the PLA part with a thickness of 2.71 mm. *Source* The authors (2020)

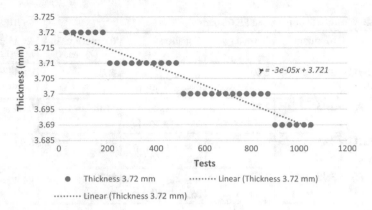

Fig. 3.6 Wear of the PLA part with a thickness of 3.72 mm. *Source* The authors (2020)

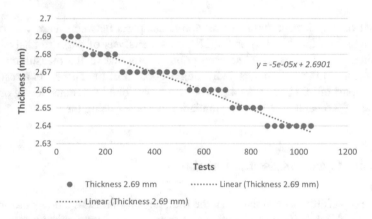

Fig. 3.7 Wear of the PETG part with a thickness of 2.69 mm. *Source* The authors (2020)

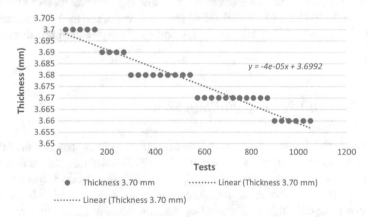

Fig. 3.8 Wear of the PETG part with a thickness of 3.70 mm. *Source* The authors (2020)

Table 3.2 Useful life of PLA and PETG parts

Thickness (mm)	Minimum (mm)	Current (y) (mm)	Equation	Result of useful life (x) (uses)	R^2 adjusted
2.69 (PETG)	2.45	2.69	$y = -5e - 05x + 2.6901$	4802	0.96
2.71 (PLA)	2.45	2.71	$y = -3e - 05x + 2.7094$	5188	0.88
3.70 (PETG)	3.45	3.7	$y = -4e - 05x + 3.6992$	4984	0.92
3.72 (PLA)	3.45	3.72	$y = -4e - 05x + 3.721$	5420	0.96

Source The authors (2020)

Based on the data, the number of equivalent parts between them was defined as essential data for the LCA and MCFA in the analysis of part use analysis. The calculation of the adjusted R^2 showed good results due to the measurement tool having errors in the millimeters, indicating a good fit of the curve. As a result, the ratio of 1 part of PETG to 1.0804 parts of PLA was obtained, considering only the thickness of 2.69 and 2.71 mm for simulation purposes using the tools.

3.5   Life Cycle Assessment of Parts

After the 3D printer printed the parts, the four steps for an LCA study were applied, being them the definition of objective and scope, the life cycle inventory analysis (LCI), the life cycle impact assessment (LCIA), and the interpretation of the analysis response (ISO 14044 2006). Steps defining goal and scope, and the LCI, are presented in Sects. 3.5.1 and 3.5.2, while the LCIA and the interpretation are discussed as results.

3.5.1   Goal and Scope Definitions

This stage of the LCA study consists of the objective, which comprises the intended application, reason, and target audience of the study. Besides this, the scope includes the definition of function, functional unit, reference flow, product system, system frontier, assumptions, allocation procedures, and method selection for characterizing the life cycle impact. Table 3.3 describes the objective, intended application, and target audience for the case study used in this book.

Table 3.3 Definition of objective and scope

Components	Features
Intended application	Compare the environmental impact of the two-part life cycle printed by the 3D printing process consisting of two different materials, PLA and PETG
Reason	Identify which of the parts has the least environmental impact
Target audience	The academic community, public, and private sectors interested in 3D printing by FDM technology printers

Source The authors (2020)

Table 3.4 Function, function unit, and reference flow

Components	Features
Function	Gap gauge
Functional unit	4.802 gap controls
Reference flow	1 part of PETG and 1.0804 part of PLA

Source The authors (2020)

3.5.1.1 Function, Functional Unit, and Reference Flow

The product function was a 3D printing part of acting as a gap gauge for the joint of doors in an automobile.

The functional unit was defined considering the reference for the use of 4.802 gap gauges, with the lowest use expected for the PETG part's disposal. This happened because, in the wear tests, the PLA had a useful life of 5.188 gap gauges for a thickness of 2.71 mm and 5.420 gap gauges for a thickness of 3.72 mm, and PETG had a useful life of 4.802 gap gauges for 2.69 mm thickness and 4.984 gap gauges for 3.70 mm thickness. From these results, the reference flow was 1 part of PETG and 1.0804 part of PLA to meet the functional unit of 4.802 gap gauges. Table 3.4 presents a summary with the definitions of function, functional unit, and the reference flow.

3.5.1.2 Definition of Products System, System Frontier, Assumptions, Allocation Procedures, Characterization Method, and LCA Simulation

The product system was the cradle to the grave. This is because the printed PETG and PLA parts were analyzed from the extraction of the raw materials that make up these materials to their final disposal.

Despite this, a data cut-off criterion of 1% of the total mass of incoming processes was defined, a value commonly applied according to Passuello et al. (2014) to exclude data of low relevance for the analysis.

As part of the system, material acquisition, preheating, and part printing processes were defined. In the material acquisition process, the inputs were PETG, PLA, and packaging, and the output was the packaging. In the preheating process, the input was energy consumption without any outputs, and in the process of printing the part, the inputs were PETG, PLA, and energy, and as outputs, the leftover PETG and PLA. As the printers evolve and their application fields increase, the number of inputs and outputs tends to increase according to the complexity of the projects.

The system boundaries of this study corresponded to materials and equipment not mentioned as part of the system. Note that it is up to the user to define how many system boundaries are essential for the printing process, such as the post-processing process. The part undergoes some sanding, polishing, or chemical application.

Concerning allocation procedures, the recommendation of ISO 14044 was considered, which says that allocation procedures should be avoided whenever it is possible (ISO 14044 2006).

Regarding the characterization method, the IMPACT 2002+ method was considered. This method was chosen because it corresponds to the method most used today to assess environmental impact in multi-indicators for products, as presented in a study by Carvalho et al. (2014). IMPACT 2002+ performs the combined assessment between the midpoint/endpoint connecting the inventory results through 14 midpoint and 4 endpoint categories (Jolliet et al. 2003).

This method covers approximately 1.500 substances as characterization factors, in addition to providing new concepts and methods of comparative evaluation between ecotoxicity effects and human toxicity (Handbook 2010). Note that the user of this methodology can adopt other LCIA methods as best suits him. However, the user must consider the processes, inputs, and outputs existing in the application under study. And also, the extension approach (impact categories and characterization factors) is addressed by the chosen method.

Finally, the LCA simulation process was carried out using the SimaPro 8.5.5 software. This software is considered the most popular and most used for product life cycle assessment. Modeling software, designed by Dutch PRé Consultants, allows systematic and transparent modeling and analysis of complex life cycles based on recommendations from the series of norms ISO 14044 (Shah et al. 2016). SimaPro has more than twenty life cycle impact assessment methods and more than nine inventory libraries providing information on thousands of products and processes (ACV BRASIL 2020).

3.5.2 Life Cycle Inventory

The LCI corresponds to the data collected and used for life cycle impact simulations. The collection of these data was carried out and constructed according to Table 3.5, which contains the relationship of materials, processes, and quantities in printing the two parts to be analyzed.

Table 3.5 Survey of materials, energies, and related processes for LCI composition of PETG and PLA parts

Process	Material	Quantity and unit
Material acquisition	PETG	1000 g
Material acquisition	PLA	1000 g
Material acquisition	PETG packing mass	75 g
Material acquisition	PLA packing mass	75 g
Material acquisition	Disposal (PETG and PLA packaging destination)	Landfill
Preheating	Equipment energy consumption for PLA	0.0013 kW h
Preheating	Energy consumption of equipment for PETG	0.0013 kW h
Part printing	Energy consumption in PLA printing	0.032 kW h
Part printing	Energy consumption in PETG printing	0.031 kW h
Part printing	Input mass of PETG	10.188 g
Part printing	Output mass of PETG	10.15 g
Part printing	PLA input mass	9.304 g
Part printing	PLA output mass	8.62 g
Part printing	Leftover PETG mass	0.09 g
Part printing	Leftover PLA mass	0.17 g
Part printing	PLA volatility	0.6838 g
Part printing	PETG volatility	0.03 g
Part printing	Disposal (destination of the leftover mass of PETG and PLA)	Landfill

Source The authors (2020)

It is noteworthy that the data used in this study corresponded to two different sources of information, being the field collection of data from the 3D printing process (real data) and data selected from the Ecoinvent 3.4 database. This database is considered the complete existing base for obtaining process and product information.

From the definition of goal and scope, and the LCI, LCIA and its interpretation were carried out.

3.6 Cost Accounting in Parts Material Flows

From the workflow described in Fig. 3.9, the sixth step was subdivided into five steps for implementing MFCA (ISO 14051 2013).

The details of these steps follow in Sects. 3.6.1 and 3.6.2.

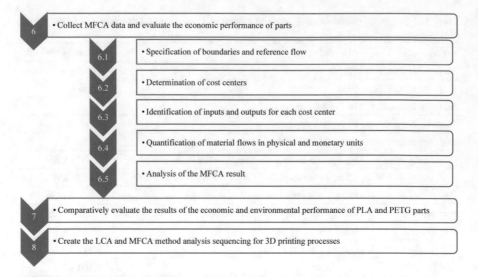

Fig. 3.9 Stages of carrying out MFCA. *Source* The authors (2020)

3.6.1   Specification of Boundaries and Analysis Reference Flow

The system to be analyzed was the printing process of the gap gauge part. According to the LCA, the boundaries were defined as the acquisition of material (raw material), the preheating of the printer, and the subsequent printing of the part. In the LCA and MFCA process, the landfill was considered as a destination for the waste generated.

Finally, as in the LCA, 1 part of the PETG and 1.0804 parts of PLA were considered reference flow for analysis by MFCA.

3.6.2   Determination of Cost Centers and Identification of Inputs and Outputs

The cost centers initially created to analyze the 3D printing process by MFCA are shown in Fig. 3.10.

As data to be used for the analysis by the MFCA, the data already collected in the LCA study of the two printed parts were initially used. Also, the costs of materials, energy, and labor were raised.

Thus, from the parameters defined in the printing, the gap gauge as printed part, and the printing tests, it was possible to measure the MFCA.

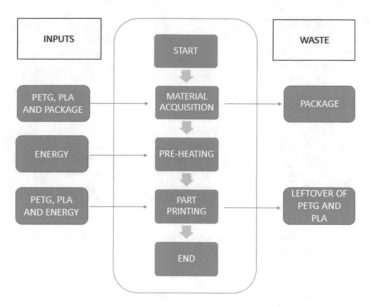

Fig. 3.10 3D printing process cost centers with identification of process inputs and outputs. *Source* The authors (2020)

3.7 LCA Results for PLA and PETG Parts

The life cycle impact assessment was obtained by comparing the life cycles of the parts produced in SimaPro v8.5.5 software. The IMPACT 2002+ method was chosen because it corresponds to the most used environmental impact assessment method in multi-indicators (Carvalho et al. 2014). Despite this, when applying this methodology, the researcher/user of the methodology can select other methods, according to their knowledge and the research in the literature. The functional unit was determined as 4.802 gap controls, and this being the value obtained for the PETG material in the thickness of 2.69 mm. This functional unit was defined given the type of product under analysis so that there is an equal comparison between the products analyzed. For the reference flow, the functional unit was used, and the calculation for the PLA was performed. As a result, 1 part of PETG was obtained for 1.0804 parts of PLA, which were used for LCA simulation. The calculation of this reference flow was based on the authors' initial choice of the PETG material, but because it is not a rule, the researcher/user can define the PLA as a functional unit for performing the calculations.

Through the IMPACT 2002+ method, 14 categories of midpoint impact were addressed: human toxicity, respiratory effects, ionizing radiation, ozone depletion, photochemical ozone formation, aquatic ecotoxicity, terrestrial ecotoxicity, aquatic acidification, aquatic eutrophication, terrestrial acidification, eutrophication, land occupation, global warming, renewable energy use, and mineral extraction.

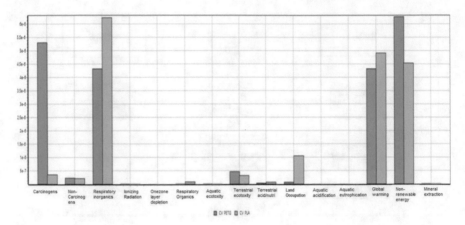

Fig. 3.11 Comparison of the life cycle scenarios of PLA and PETG by the IMPACT 2002+ method (normalization). *Source* The authors (2020)

The normalization graphs are obtained in the results. This is because the focus of the LCA application is a decision support. In this sense, standardization makes it possible to verify the intensity of the potential impact contribution of each category. The results were also used in the form of a single score (Ecopoint) to visualize the aggregate environmental impact potential for each of the filaments used in the printed parts under study.

Through the normalization of the impact categories (Fig. 3.11), it was possible to realize that the four categories that most influenced the impact potential for LC of PETG were, in descending order: non-renewable energies, carcinogens, inorganic respiratory, and terrestrial ecotoxicity. For the LC of PLA, the categories were, in descending order: inorganic respiratory, global warming, non-renewable energy, and land occupation. The other categories presented contributions below 2% to the overall impact potential.

In this context, it is perceived that the most significant impact potentials of the analyzed LCs are in five impact categories, as already described, thus defining the focus of analysis to identify the causes that influence their environmental performance.

Another way of visualizing the results of multi-indicators for the two LCs is from the Ecopoint (Pt). In this study, it can be seen when visually comparing Fig. 3.12 that the differences in life cycles showed a total μPt of 17.9 for the LC of PLA and 21.1 for the LC of PETG. This result suggests that in the simulation performed, for the 3D printing scenario defined by FDM technology, the part printed in PLA presented a lower potential for global environmental impact than the part in PETG.

Despite this result, the LC of PLA presented an environmental impact potential for the inorganic respiratory category of 6.50 μPt against 4.95 μPt of PETG. This category of impact is considered one of the most significant within the IMPACT 2002+ method, presenting a relatively higher value of normalization factor within damage to human health than the other categories in this class (Goedkoop and Spriesnma

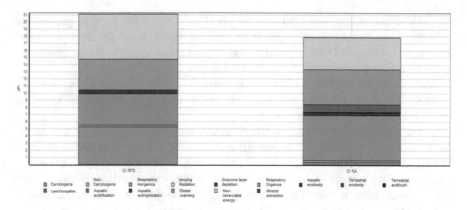

Fig. 3.12 Comparison of the life cycle scenarios of PLA and PETG by the IMPACT 2002+ method (Ecopoint). *Source* The authors (2020)

2000). It gives importance to the damage to human health. Still, it is related to the emission of particulate material <2.5 μm, which has been linked to several types of cancer, and a variety of non-carcinogenicity in people (Spadaro and Rabl 1999), which, in addition to the gaseous compounds, inorganic, such as carbon monoxide (CO), nitrogen oxides (NO_x), and oxides of sulfur (SO_x), much of which are toxic to humans and can cause lung damage and also inflammation, especially in children, the elderly, and people with asthma (Dibofori-Orji and Braides 2013; World Health Organization 2005).

To better understand and explain the results of the impact categories that obtained the highest values for the LCs of PLA and PETG, the results of the impact contributions of the processes involved in these polymers' life cycles were analyzed. Also, the contributions of the materials considered in these processes throughout the life cycle of each filament were investigated.

In this context, Sect. 3.7.1 dealing with inventory results (materials contribution) and Sect. 3.7.2 dealing with process contributions results throughout the life cycle of 3D printing processes with PLA and PETG were elaborated. For the analysis of these results, a cut of 1% of the data presented in the graphs and tables was performed to facilitate the results' visualization and interpretation.

3.7.1 Inventory Results (Materials Contribution)

In this section, the potential impact contribution of materials present in the life cycles analyzed for 3D printing in PLA and PETG is presented. According to the previous section, the results are presented by composing a graph, and a table for each impact category considered relevant, that is, with the highest impact potential values. The impact categories analyzed were global warming, carcinogens, inorganic respiratory, non-renewable energy, and land occupation. The tables were used to present the material names, numerical information (in micro-Ecopoint), and the identification of the materials presented by the SimaPro software, corresponding to the nomenclature of the materials present in the Ecoinvent 3.4 database.

Thus, Table 3.6 and Fig. 3.13 present the inventory results for the global warming category. The total emissions for the LCs of PETG and PLA were 0.0428 kgeqCO$_2$ and 0.0457 kgeqCO$_2$, respectively.

Composing these total values, the largest contributions related to the LC of the PETG were carbon dioxide (fossil), representing a total of 88.90%; methane (fossil) (4.49%); methane (biogenic) (2.88%), and carbon dioxide by soil transformation (2.18%). For the PLA LC, the main contributors were carbon dioxide (fossil) (86.77%); dinitrogen monoxide (4.08%); carbon dioxide, by soil transformation (3.13%); methane of fossil origin (2.92%), and methane (biogenic) (2.60%).

It was found that carbon dioxide (fossil), methane (fossil), carbon dioxide (soil transformation), and methane (biogenic) are common to the two life cycles analyzed. The contribution of carbon dioxide (fossil) of the PETG material refers to about 1.2% of the contribution generated by the PLA material. The contribution of methane (fossil) of PETG refers to about 21.18% of PLA's contribution. The contribution of

Materials	LC PETG	LC PLA
Carbon dioxide, fossil	3.8438	4.2700
Carbon dioxide, land transformation	0.0941	0.1542
Dinitrogen monoxide	0.0497	0.2007
Methane, biogenic	0.1245	0.1280
Methane, fossil	0.1942	0.1439

Table 3.6 Inventory results of the global warming category

Source The authors (2020)

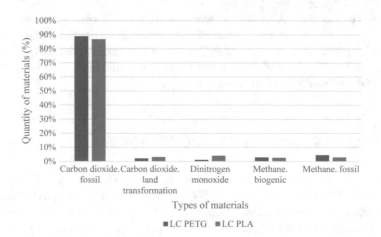

Fig. 3.13 Inventory results for the global warming category. *Source* The authors (2020)

methane (biogenic) of the PETG material refers to about 6.65% of PLA's contribution. On the other hand, the contribution of carbon dioxide (soil transformation) of PLA material refers to 17.89% of the contribution generated by PETG material. Studies such as that of Lamnatou et al. (2019) obtained higher impact results in the global warming category for PETG, while Madival et al. (2009) and Vink et al. (2003) showed this greater impact on global warming for PLA.

Table 3.7 and Fig. 3.14 present the inventory results for the carcinogens category. The total emission results of carcinogens were 0.0134 kgeqC2H3Cl and 0.000845 kgeqC2H3Cl for PETG and PLA, respectively. The largest contributions related to the LC of the PETG were aromatic hydrocarbons (air), representing a total of 99.20%. For the PLA LC, the main contributors were aromatic hydrocarbons (air) (80.91%); benzo(a)pyrene (3.75%); arsenic (water) (3.20%); atrazine

Table 3.7 Inventory results of the carcinogens category

Materials	LC PETG	LC PLA
Arsenic (air)	0.0061	0.0060
Arsenic (water)	0.0084	0.0114
Atrazine	0.0000	0.0112
Benzo(a)pyrene	0.0086	0.0134
Dioxin, 2,3,7,8 tetrachlorodibenzo-*p*-dioxin	0.0060	0.0101
Hydrocarbons, aromatic	5.2666	0.2893
PAH, polycyclic aromatic hydrocarbons	0.0050	0.0094

Source The authors (2020)

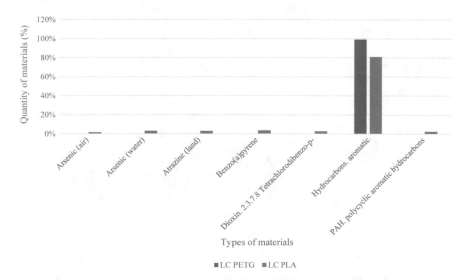

Fig. 3.14 Inventory results of the carcinogens category. *Source* The authors (2020)

(3.15% Pt); 2,3,7,8-tetrachlorodibenzo-*p*-dioxin (2.2%); HPAs-polycyclic aromatic hydrocarbons (2.63%).

It was found that only aromatic hydrocarbons (air) are common to the two life cycles considered. Despite this, due to the much higher value of aromatic hydrocarbon emissions in the LC of the PETG, about 0.0133 kgeqC2H3Cl against 0.000683 kgeqC2H3CL of the PLA, made its life cycle present a significant impact potential in relation to the PLA. According to Silva and Kulay (2006), air chemical pollutants such as aromatic hydrocarbons are related to the development of many lesions in humans, presenting significant effects associated with cancerous problems.

Table 3.8 and Fig. 3.15 present the inventory results for the category inorganic respiratory. The total emissions shown by the LCs of the PETG and PLA for this category were 4.39.10–5 kgeqPM2.5 and 5.97.10–5 kgeqPM2.5, respectively. The most considerable contributions related to PETG LC were particulate material <2.5 μm (PM2.5), representing a total of 55.13%; nitrogen oxides (22.58%) and sulfur dioxide (21.79%).

For the LC of PLA, the main contributors were particulate material <2.5 μm, representing a total of 59.49%; nitrogen oxides (18.99%); sulfur dioxide (16.46%); and ammonia (5.06%). The PM2.5 contribution of PLA material refers to about 3.8% of the contribution generated by PETG material. The contribution of PETG material

Table 3.8 Inventory results of the respiratory inorganics category	Materials	LC PETG	LC PLA
	Ammonia	0.0215	0.3154
	Nitrogen oxides	0.9788	1.1830
	Particulates, <2.5 μm	2.3893	3.7061
	Sulfur dioxide	0.9444	1.0254

Source The authors (2020)

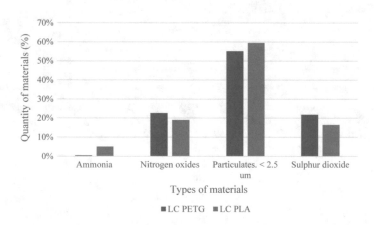

Fig. 3.15 Inventory results of the respiratory inorganics category. *Source* The authors (2020)

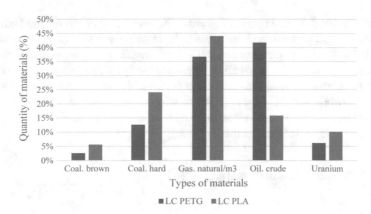

Fig. 3.16 Inventory results of the non-renewable energy category. *Source* The authors (2020)

Table 3.9 Inventory results of the non-renewable energy category	Materials	LC PETG	LC PLA
	Coal, brown	0.1603	0.2526
	Coal, hard	0.7885	1.0914
	Gas, natural (m³)	2.3048	1.9985
	Oil, crude	2.6193	0.7173
	Uranium	0.3863	0.4594

Source The authors (2020)

related to sulfur dioxide and nitrogen oxides refers to approximately 13.93% and 8.63% of the PLA's contribution, respectively.

Figure 3.16 and Table 3.9 describe the five substances that make up the inventory results for non-renewable energy. The total impact potentials for LCs of PETG and PLA due to the consumption of non-renewable materials used in energy production were 0.953 MJprimary and 0.640 MJprimary, respectively.

For the PETG LC, there was a predominance of crude oil and natural gas, which together corresponded to 78.50% of the total substances. For the LC of PLA, natural gas represented 44.02%, followed by hard coal with 24.04% contribution.

When the materials were analyzed in comparison, the PLA presented representativeness of 44.02% while the PETG 36.74% to the substance natural gas. This value grows when the substances crude oil and hard coal are compared. For hard coal, PLA presented an inventory of 24.04%, while for crude oil, PETG represented 41.7%. It is observed that the compounds belonging to this impact category, usually related to oil, coal, and natural gas consumption, correlate with the emission compounds present in the global warming category. Therefore, the proximity of the results of these two categories can be perceived.

Finally, Table 3.10 and Fig. 3.17 present the inventory results for the land occupation category. The total land use areas calculated for the LCs of PETG and PLA were 0.000918 m^2 organic plow and 0.0123 m^2 organic plow, respectively.

For this inventory, the contributing substances are grouped into a particular category. PLA material presents the occupation per annual crop corresponding to 84.32% of the total, while for PETG, this representation corresponds to 0.20%. The intensive forest occupation is equivalent to 68.52% of the total, while for the PLA, this significance holds 5.16% of the total.

Table 3.10 Inventory results of the land occupation category

Materials	LC PETG	LC PLA
Remaining substances	0.0011	0.0012
Occupation, annual crop	0.0001	0.8919
Occupation, annual crop, non-irrigated, intensive	0.0024	0.0533
Occupation, dump site	0.0057	0.0079
Occupation, forest, intensive	0.0501	0.0546
Occupation, grassland, natural (non-use)	0.0008	0.0011
Occupation, industrial area	0.0040	0.0312
Occupation, mineral extraction site	0.0011	0.0016
Occupation, traffic area, rail/road embankment	0.0038	0.0047
Occupation, traffic area, road network	0.0036	0.0037

Source The authors (2020)

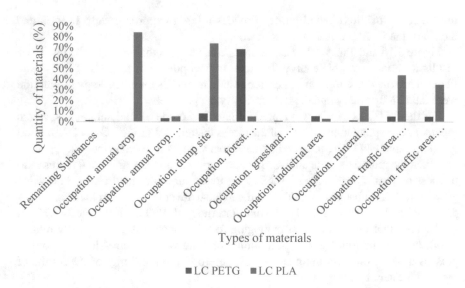

Fig. 3.17 Inventory results of the land occupation category. *Source* The authors (2020)

In this context, to make it possible to conclude the ratios of the calculated values of impact potential for the five categories analyzed in this topic, the process's contribution analysis was carried out, which follows Sect. 3.7.2.

3.7.2 Process Contribution

As well as analyzed by materials, the five categories of greatest impact were analyzed for the contribution by the process. In this section, the nomenclature of the processes was used, as provided in the simulation results in the SimaPro software, to ensure fidelity to the database. In this context, a graph and a table were prepared for each impact category analyzed, and in Sect. 3.7.1, to facilitate the visualization of the results and the nomenclatures of the processes. Thus, carcinogens presented 15 contributions per process, and those considered, given the data cut (1%), as follows Fig. 3.18.

It can be seen that natural gas production represented 68.82% of the total contribution to PLA, while for PETG, it represented 3.58% of the contribution. This natural gas present in PLA originates from two main processes being the energy consumption in the printing process and the heat consumption in the polymer extrusion process for the production of PLA filament. In PETG, this natural gas is consumed significantly in correlated processes of PLA. Despite this, its percentage value is much lower due

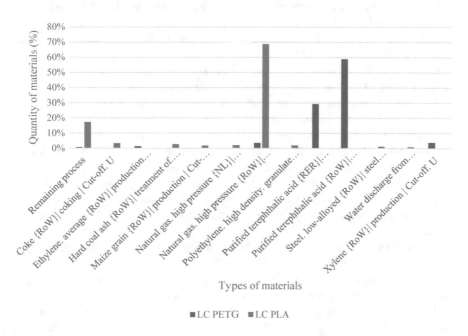

Fig. 3.18 Process contribution of the carcinogens category. *Source* The authors (2020)

Table 3.11 Process contribution of the carcinogens category

Process	LC PETG	LC PLA
Remaining process	0.0380	0.0618
Coke {RoW}	coking	Cut-off, U	0.0070	0.0118
Ethylene, average {RoW}	production	Cut-off, U	0.0744	0.0001
Hard coal ash {RoW}	treatment of, residual material landfill	Cut-off, U	0.0062	0.0095
Maize grain {RoW}	production	Cut-off, U	0.0000	0.0063
Natural gas, high pressure {NL}	petroleum and gas production, on-shore	Cut-off, U	0.0021	0.0072
Natural gas, high pressure {RoW}	natural gas production	Cut-off, U	0.1902	0.2461
Polyethylene, high density, granulate {RoW}	production	Cut-off, U	0.0006	0.0070
Purified terephthalic acid {RER}	production	Cut-off, U	1.5558	0.0000
Purified terephthalic acid {RoW}	production	Cut-off, U	3.1294	0.0000
Steel, low-alloyed {RoW}	steel production, electric, low-alloyed	Cut-off, U	0.0045	0.0044
Xylene {RoW}	production	Cut-off, U	0.0031	0.0033

Source The authors (2020)

to the greater impact potential in this category by terephthalic acid. This purified acid production represented 88.65% for PETG, and there was no contribution when considered for PLA. It corresponds to one of the main reagents for PET production, along with ethylene glycol (Jou et al. 2015). It is related to the phase of the manufacturing life cycle of the PET chemical that will then be used in the manufacture of the PETG filament (Table 3.11).

Global warming presented a total of 148 contributions from the process, and with the data cut (1%), 18 contributions were considered (Fig. 3.19).

The production of high voltage electricity represented 35.95% of contribution to the PLA, while for the PETG, this amount was 27.58% when added in both cases. According to the simulation results, this high voltage energy is linked to energy consumption, both from the printer and in the manufacture of each material's filaments. Despite this, there was a significant difference compared to xylene production, with 16.80% for PETG material and no contribution to PLA (Table 3.12). Xylene is an organic solvent used in the purification of terephthalic acid. This has considerable global warming potential due to the emission of gases such as CO_2 in its production process.

For land occupation, 38 factors were generated, of which 27 contributions per process were considered due to the 1% cut (Fig. 3.20).

The total impact potentials in land occupation for PETG and PLA were 0.000918 m^2 organic plow and 0.0123 m^2 organic plow, respectively.

The PETG presented a contribution of 32.48% when the process is related to wood for milling, forestry, and sustainable forest management, while for the PLA, this value is 0.16% (Table 3.13). As traced in the process network of the PETG LC, it

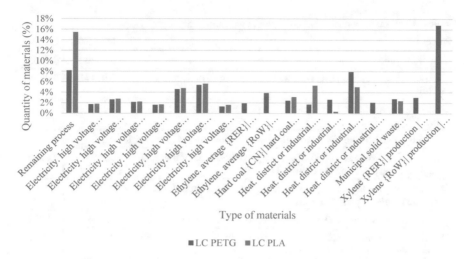

Fig. 3.19 Process contribution of the global warming category with a cut above 1%. *Source* The authors (2020)

was verified that this consumption of wood is related to two main processes. Firstly, the manufacture of paper packaging (paper box) used for packaging PETG. Second and the manufacture of wooden pallets used in the transport of raw materials in this polymer's manufacture.

These processes of paper packaging and wood pallet manufacturing also appear in the LC of PLA. Still, due to the origin of the polylactic acid being of vegetable source, it was perceived as a significant contribution of the corn grain production process, representing 88.86% of this acid's impact potential. At the same time, PETG had no contribution because it was of fossil source. In this context, it can be stated that the greatest potential impact on the part of the PLA was due to the consumption of area for vegetable planting for subsequent manufacture of acid, a fact that does not exist in the LC of PETG.

Figure 3.21 shows the contribution of processes to the non-renewable energy category, where 54 contributions were obtained, but only 27 were considered for constructing the graph (1% cut).

It was observed that the LC of the PETG presented a total impact potential in this category of 0.953 MJprimary, while the LC of the PLA presented as a result of 0.640 MJprimary.

Table 3.14 shows the contribution of xylene to PETG, with a representation of 40.54%, while for PLA, there was no contribution to this process. As already mentioned for the global warming category, xylene is a solvent used to purify terephthalic acid. Its significant impact on the non-renewable energy category was due to fossil energy sources' use to produce this substance. Fossil fuels are burned for the production of this solvent, characterizing impact on non-renewable energy consumption. Consequently, greenhouse gases are released into the atmosphere, impacting the global warming category.

Table 3.12 Process contribution of the global warming category

Process	LC PETG	LC PLA
Remaining process	0.3529	0.7603
Electricity, high voltage {BR} I electricity production, hard coal I Cut-off, U	0.0744	0.0884
Electricity, high voltage {BR} I electricity production, hydro, reservoir, tropical region I Cut-off, U	0.1140	0.1354
Electricity, high voltage {BR} I electricity production, lignite I Cut-off, U	0.0927	0.1101
Electricity, high voltage {BR} I electricity production, natural gas, combined cycle power plant I Cut-off, U	0.0698	0.0829
Electricity, high voltage {BR} I electricity production, natural gas, conventional power plant I Cut-off, U	0.1987	0.2361
Electricity, high voltage {BR} I electricity production, oil I Cut-off, U	0.2339	0.2779
Electricity, high voltage {RoW} I electricity production, natural gas, conventional power plant I Cut-off, U	0.0561	0.0781
Ethylene, average {RER} I production I Cut-off, U	0.0825	0.0001
Ethylene, average {RoW} I production I Cut-off, U	0.1671	0.0002
Hard coal {CN} I hard coal mine operation and hard coal preparation I Cut-off, U	0.1048	0.1536
Heat, district or industrial, natural gas {Europe without Switzerland} I heat production, natural gas, at industrial furnace > 100 kW I Cut-off, U	0.0735	0.2608
Heat, district or industrial, natural gas {RoW} I heat production, natural gas, at industrial furnace >100 kW I Cut-off, U	0.1136	0.0174
Heat, district or industrial, other than natural gas {RoW} I heat production, at hard coal industrial furnace 1–10 mW I Cut-off, U	0.3416	0.2478
Heat, district or industrial, other than natural gas {RoW} I heat production, light fuel oil, at industrial furnace 1 mW I Cut-off, U	0.0902	0.0053
Municipal solid waste {RoW} I treatment of, sanitary landfill I Cut-off, U	0.1199	0.1186
Xylene {RER} I production I Cut-off, U	0.1312	0.0000
Xylene {RoW} I production I Cut-off, U	0.7266	0.0001

Source The authors (2020)

Another perceived result refers to the coal mine operation and coal preparation, representing 20.32% for PLA and 9.69% for PETG. This coal is related to the use of thermal energy in the production of PLA and PETG polymers. Still, it is also related to the drying process of corn, this being the raw material for the manufacture of PLA. Due to this increase in use in the drying of corn grains, PLA presented a greater contribution to this coal process than PETG.

Besides, there was also the contribution of natural gas production, which represented 34.97% for PLA while for PETG only 12.30%. However, these processes reflect 62.53% for PETG and 55.29% of the total contribution when added together. As already mentioned in the global warming category, natural gas production is

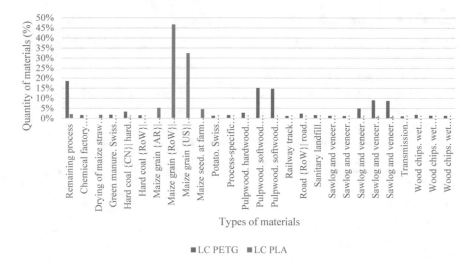

Fig. 3.20 Process contribution of the land occupation category with a cut above 1%. *Source* The authors (2020)

related to electricity production, both for consumption in the printing process and for consumption in the manufacture of PLA and PETG filaments.

Finally, inorganic respiration presented 140 contributions per process, but 21 of these presented values greater than 1% of this category's total impact contribution (Fig. 3.22). The total impact potentials for the PETG and PLA LCs presented results of 4.39×10^{-5} kgeqPM2.5 and 5.97×10^{-5} kgeqPM2.5, respectively.

In this scenario, electricity production was responsible for 43.03% for the LC of the PLA, being a close value when compared to the PETG, which obtained 41.66%. Corn grain production again showed a greater contribution to PLA, 4.62%, while there was no contribution to PETG material. That is, the processes already evidenced in the categories of the impact of global warming and non-renewable energy, in general, contributed significantly to the impact potential of this category of inorganic respiration. For this reason, the contributions of these three mentioned categories come from correlated processes. They present similar intensities of environmental impact potential in the simulations of the LCs of PETG and PLA (Table 3.15).

The sequence of results of this study is given through Sect. 3.8, which presents the results obtained by applying the MFCA tool for measuring the economic performance of material and energy flows PLA and PETG materials. Subsequently, in Sect. 3.9, the results of both tools are discussed together to conclude the economic-environmental performance of the analyzed filaments.

Table 3.13 Process contribution of the land occupation category

Process	LC PETG	LC PLA
Remaining process	0.0135	0.0217
Chemical factory, organics {RoW} I construction I Cut-off, U	0.0012	0.0006
Drying of maize straw and whole-plant {RoW} I processing I Cut-off, U	0.0000	0.0187
Green manure, Swiss integrated production, until March {CH} I production I Cut-off, U	0.0013	0.0005
Hard coal {CN} I hard coal mine operation and hard coal preparation I Cut-off, U	0.0024	0.0036
Hard coal {RoW} I hard coal mine operation and hard coal preparation I Cut-off, U	0.0011	0.0018
Maize grain {AR} I maize grain production I Cut-off, U	0.0000	0.0548
Maize grain {RoW} I production I Cut-off, U	0.0000	0.4941
Maize grain {US} I production I Cut-off, U	0.0000	0.3427
Maize seed, at farm {GLO} I production I Cut-off, U	0.0000	0.0482
Potato, Swiss integrated production {CH} I potato production, Swiss integrated production, intensive I Cut-off, U	0.0009	0.0003
Process-specific burden, sanitary landfill {RoW} I processing I Cut-off, U	0.0011	0.0011
Pulpwood, hardwood, measured as solid wood under bark {RoW} I hardwood forestry, birch, sustainable forest management I Cut-off, U	0.0020	0.0008
Pulpwood, softwood, measured as solid wood under bark {RoW} I softwood forestry, pine, sustainable forest management I Cut-off, U	0.0110	0.0005
Pulpwood, softwood, measured as solid wood under bark {RoW} I softwood forestry, spruce, sustainable forest management I Cut-off, U	0.0107	0.0005
Railway track {RoW} I construction I Cut-off, U	0.0008	0.0010
Road {RoW} I road construction I Cut-off, U	0.0017	0.0014
Sanitary landfill facility {RoW} I construction I Cut-off, U	0.0012	0.0012
Sawlog and veneer log, hardwood, measured as solid wood under bark {DE} I hardwood forestry, beech, sustainable forest management I Cut-off, U	0.0010	0.0014
Sawlog and veneer log, hardwood, measured as solid wood under bark {RoW} I hardwood forestry, beech, sustainable forest management I Cut-off, U	0.0009	0.0009
Sawlog and veneer log, softwood, measured as solid wood under bark {CA-QC} I softwood forestry, mixed species, boreal forest I Cut-off, U	0.0036	0.0037
Sawlog and veneer log, softwood, measured as solid wood under bark {RoW} I softwood forestry, pine, sustainable forest management I Cut-off, U	0.0066	0.0071
Sawlog and veneer log, softwood, measured as solid wood under bark {RoW} I softwood forestry, spruce, sustainable forest management I Cut-off, U	0.0064	0.0068

(continued)

Table 3.13 (continued)

Process	LC PETG	LC PLA
Transmission network, electricity, high voltage {CA-QC} I transmission network construction, electricity, high voltage I Cut-off, U	0.0008	0.0011
Wood chips, wet, measured as dry mass {RoW} I hardwood forestry, birch, sustainable forest management I Cut-off, U	0.0014	0.0016
Wood chips, wet, measured as dry mass {RoW} I softwood forestry, pine, sustainable forest management I Cut-off, U	0.0011	0.0012
Wood chips, wet, measured as dry mass {RoW} I softwood forestry, spruce, sustainable forest management I Cut-off, U	0.0010	0.0011

Source The authors (2020)

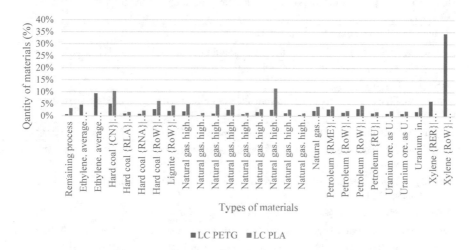

Fig. 3.21 Process contribution of the non-renewable energy category with a cut above 1%. *Source* The authors (2020)

3.8 Results of the MFCA for PLA and PETG

From the definition of the cost centers and the collection of input and output data, Tables 3.16 and 3.17 were obtained. These tables show the CC for the printing process and the number of people involved in each CC, and the time spent carrying out the activity. The defined CC were CC1—material acquisition; CC2—preheating; CC3—part printing.

The quantity of PLA and PETG was obtained through the material purchasing cost. The burr masses and material volatility determined the wastes. Volatility was calculated by the difference between the total mass (piece + burr) and the mass of the filament used.

The cost of the system (labor) was defined as R$20.00 per hour of printing. This cost was obtained using the average price of three different printing service

Table 3.14 Process contribution of the non-renewable energy category

Process	LC PETG	LC PLA
Remaining process	0.0402	0.1452
Ethylene, average {RER} I production I Cut-off, U	0.2902	0.0004
Ethylene, average {RoW} I production I Cut-off, U	0.5876	0.0008
Hard coal {CN} I hard coal mine operation and hard coal preparation I Cut-off, U	0.3195	0.4680
Hard coal {RLA} I hard coal mine operation and hard coal preparation I Cut-off, U	0.0614	0.0729
Hard coal {RNA} I hard coal mine operation and hard coal preparation I Cut-off, U	0.0520	0.0991
Hard coal {RoW} I hard coal mine operation and hard coal preparation I Cut-off, U	0.1749	0.2829
Lignite {RoW} I mine operation I Cut-off, U	0.1267	0.1973
Natural gas, high pressure {DZ} I natural gas production I Cut-off, U	0.1188	0.2232
Natural gas, high pressure {NL} I petroleum and gas production, on-shore I Cut-off, U	0.0166	0.0554
Natural gas, high pressure {NO} I petroleum and gas production, off-shore I Cut-off, U	0.0631	0.2188
Natural gas, high pressure {RoW} I natural gas production I Cut-off, U	0.1571	0.2032
Natural gas, high pressure {RoW} I petroleum and gas production, off-shore I Cut-off, U	0.0458	0.0589
Natural gas, high pressure {RoW} I petroleum and gas production, on-shore I Cut-off, U	0.1024	0.1317
Natural gas, high pressure {RU} I natural gas production I Cut-off, U	0.1643	0.5218
Natural gas, high pressure {US} I natural gas production I Cut-off, U	0.0737	0.1239
Natural gas, high pressure {US} I petroleum and gas production, on-shore I Cut-off, U	0.0301	0.0507
Natural gas, unprocessed, at extraction {GLO} I production I Cut-off, U	0.1387	0.1802
Petroleum {RME} I production, on-shore I Cut-off, U	0.1797	0.1901
Petroleum {RoW} I petroleum and gas production, off-shore I Cut-off, U	0.0954	0.1008
Petroleum {RoW} I petroleum and gas production, on-shore I Cut-off, U	0.1918	0.2029
Petroleum {RU} I production, on-shore I Cut-off, U	0.0785	0.0830
Uranium ore, as U {RNA} I uranium mine operation, underground I Cut-off, U	0.0689	0.0991
Uranium ore, as U {RoW} I uranium mine operation, underground I Cut-off, U	0.0651	0.0936
Uranium, in yellowcake {GLO} I uranium production, in yellowcake, in-situ leaching I Cut-off, U	0.1179	0.1696
Xylene {RER} I production I Cut-off, U	0.3889	0.0000

(continued)

Table 3.14 (continued)

Process	LC PETG	LC PLA
Xylene {RoW} I production I Cut-off, U	2.1541	0.0002

Source The authors (2020)

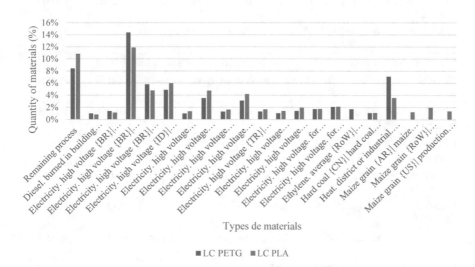

Types de materials

■ LC PETG ■ LC PLA

Fig. 3.22 Process contribution of the inorganic respiratory category with a cut above 1%. *Source* The authors (2020)

suppliers. The PLA part printing lasted 3.112 s plus 130 s of preheating of the printer, while the PETG lasted 46.57 min plus 2 min of preheating. The energy calculation considered the time and power (40 W) to obtain the kWh. The cost of kWh was R$0.77, including ICMS and PIS/COFINS fees (Brazilian fees), based on the current tariff for residences of the COPEL (local electric power concessionaire). The waste management was defined based on the estimated quantity of material generated per day (kg), the number of days in the year (365 days), and the waste collection fee (R$471.60).

Based on the separation of costs between products and waste, a comparison was made for the PLA and PETG materials. The following nomenclature was used to compare costs:

- CIM—Cost of input material;
- CE—Cost of energy;
- CS—Cost of system (labor);
- CW—Cost of waste;
- CMP—Cost of material (product);
- CMW—Cost of material (waste);
- TCP—Total cost of product;
- TCW—Total cost of waste.

Table 3.15 Process contribution of the inorganic respiratory category

Process	LC PETG	LC PLA
Remaining process	0.3654	0.6754
Diesel, burned in building machine {GLO} I processing I Cut-off, U	0.0434	0.0507
Electricity, high voltage {BR} I electricity production, hard coal I Cut-off, U	0.0600	0.0713
Electricity, high voltage {BR} I electricity production, lignite I Cut-off, U	0.6231	0.7403
Electricity, high voltage {BR} I electricity production, oil I Cut-off, U	0.2508	0.2979
Electricity, high voltage {ID} I electricity production, lignite I Cut-off, U	0.2117	0.3719
Electricity, high voltage {MRO, US only} I electricity production, lignite I Cut-off, U	0.0430	0.0838
Electricity, high voltage {RFC} I electricity production, lignite I Cut-off, U	0.1528	0.2969
Electricity, high voltage {RU} I heat and power co-generation, lignite I Cut-off, U	0.0562	0.1001
Electricity, high voltage {SERC} I electricity production, lignite I Cut-off, U	0.1345	0.2610
Electricity, high voltage {TR} I electricity production, lignite I Cut-off, U	0.0571	0.1043
Electricity, high voltage {TRE} I electricity production, lignite I Cut-off, U	0.0451	0.0877
Electricity, high voltage {WECC, US only} I electricity production, lignite I Cut-off, U	0.0616	0.1223
Electricity, high voltage, for internal use in coal mining {CN} I electricity production, hard coal, at coal mine power plant I Cut-off, U	0.0748	0.1096
Electricity, high voltage, for internal use in coal mining {RoW} I electricity production, hard coal, at coal mine power plant I Cut-off, U	0.0910	0.1333
Ethylene, average {RoW} I production I Cut-off, U	0.0750	0.0001
Hard coal {CN} I hard coal mine operation and hard coal preparation I Cut-off, U	0.0475	0.0696
Heat, district or industrial, other than natural gas {RoW} I heat production, at hard coal industrial furnace 1–10 MW I Cut-off, U	0.3068	0.2225
Maize grain {AR} I maize grain production I Cut-off, U	0.0000	0.0768
Maize grain {RoW} I production I Cut-off, U	0.0000	0.1246
Maize grain {US} I production I Cut-off, U	0.0000	0.0864

Source The authors (2020)

Table 3.16 MFCA data for the PETG

CC	Material	Input (g)	Output/product (g)	Output/waste (g)	Cost (R$/g)	Destination cost (R$/g)	Energy (kWh)	Labor	
CC1	PETG	10.19	9.3038	0	0.16	0	0	0	
	Packing	0.764	0	0.764	0	0.206	0	0	
CC2	0	0	0	0	0	0	0.0013	No	1
								Cost (R$/s)	0.0055
								Time (s)	120
CC3	PETG	10.19	10.06	0.128	0.16	0.206	0.031	No	1
								Cost (R$/s)	0.0055
								Time (s)	32,794

Source The authors (2020)

Table 3.17 MFCA data for the PLA

CC	Material	Input (g)	Output/product (g)	Output/waste (g)	Cost (R$/g)	Destination cost (R$/g)	Energy (kWh)	Labor		
CC1	PLA	10.052	0	0	0.105	0	0	No	0	1
	Packing	0.754	0	0.754	0	0.206	0		Cost (R$/s)	0.0055
									Time (s)	130
CC2	0	0	0	s0	0	0	0.0014	No	0	1
									Cost (R$/min)	0.0055
CC3	PLA	10.052	9.13	0.9227	0.105	0.206	0.034		Time (s)	3,112

Source The authors (2020)

Figure 3.23 represents the comparison to the cost of product for CC1 (material acquisition). It is observed that all costs refer to the product for both materials. In this cost center, there was no consumption of energy or labor.

In this comparison, PETG showed higher product cost, given that this cost is related to the price of the PETG, which is higher than PLA. Besides this, the mass of PETG used in the printing process was higher than the PLA mass.

In CC2 (preheating), as there were no mass flows, the result was null for both materials.

For CC3 (part printing), the material cost, energy cost, and system cost (labor) are represented by Fig. 3.24, relative to the contribution to the product.

In this process, PETG showed a cost higher than PLA because of the same observations made in the CC1. For the energy cost, PETG has a slightly higher cost, but

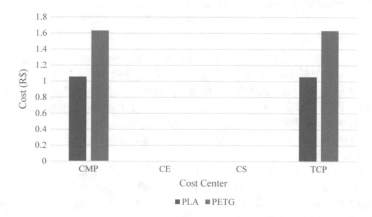

Fig. 3.23 Comparison of product cost between PLA and PETG for the CC1. *Source* The authors (2020)

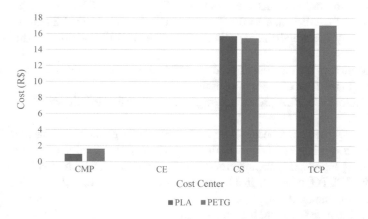

Fig. 3.24 Comparison of product cost between PLA and PETG for the CC3. *Source* The authors (2020)

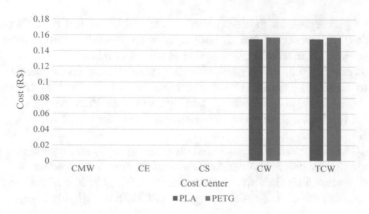

Fig. 3.25 Comparison of waste cost between PLA and PETG for the CC1. *Source* The authors (2020)

it is indistinguishable in Fig. 3.24 due to its order of magnitude. The results for this energy cost were R$0.023762 for PETG and R$0.023747 for PLA. This small difference was due to the time difference in the printing process between the materials (3.112 s for PLA and 2.794 s for PETG). Hence, the mass incorporated in the print part was higher for the PETG (10.06 g) than for the PLA (9.13 g). This was also reflected in the cost of system, in which PETG cost was again higher due to the mass incorporated in the print part.

For the cost of waste, the results are shown in Fig. 3.25. It shows the cost of waste for the CC1 (material acquisition), being that the PETG presented the higher cost.

Figure 3.25 shows that the value of the cost of waste obtained a contribution only from the cost of waste management (CWM), presenting a result equal to the total cost of waste (TCW). Although there was no loss of filaments in the CC1, there was a loss of filament packaging, which was considered a destination for the landfill.

It was also observed that the result of PETG for this TCW was slightly higher than that of PLA because as the value of mass consumed by PETG was higher. Consequently, the proportional packaging mass to be discarded was also higher than PLA.

As mentioned for the cost of product, CC2 presented a null result for the cost of waste due to the lack of mass flow in this CC.

Finally, for CC3, PLA presented a higher cost of waste in all the variables involved (Fig. 3.26).

This result happened because the PLA presented a greater loss mass (0.9227 g) than the loss of PETG (0.128 g). In this context, despite the value of the unit cost of PETG being 52.38% higher than the PLA, the loss mass of PLA was 721% higher than that of PETG, resulting in a higher cost of waste for the PLA.

Thus, from the LCA and MFCA results, Sect. 3.9 was elaborated, which addresses the comparative assessment between the materials.

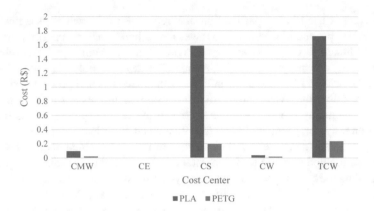

Fig. 3.26 Comparison of waste cost between PLA and PETG for the CC3. *Source* The authors (2020)

3.9 Comparative Financial-Environmental Performance Evaluation of the PLA and PETG Print Parts

The comparative evaluation of the financial-environmental performance was obtained using the results found with the LCA and the MFCA.

First, from the results presented in Sect. 3.7, referring to the total environmental impact potentials of the PETG and PLA life cycles, it was noticed that the first one presented worse environmental performance than the PLA. The categories of non-renewable energy, carcinogens, respiratory inorganics, and global warming had the most contributed to the worst performance of PETG (Table 3.18). For the PLA, the categories of respiratory inorganics, global warming, and non-renewable energy were the most significant to contribute to this filament's environmental performance.

Regarding the results in Sects. 3.7.1 and 3.7.2, PETG obtained a greater contribution to the category of non-renewable energy due to the consumption of crude oil, natural gas, and mineral coal. Crude oil was identified in the energy production processes, both thermal and electrical, and mainly in xylene's production process.

Table 3.18 Results of the environmental impact in Pt (Ecopoint)

Impact category	Unity	LC PETG	LC PLA
Total	μPt	21.1	17.9
Carcinogens	μPt	5.31	0.358
Respiratory inorganic	μPt	4.33	6.23
Land occupation	μPt	0.0731	1.06
Global warning	μPt	4.32	4.92
Non-renewable energy	μPt	6.27	4.54

Source The authors (2020)

This solvent is used in the purification of terephthalic acid, which is a raw material for PETG production. Mineral coal was found in the thermal energy production processes linked to the manufacture of PETG. Finally, natural gas was related to electric energy consumption, both in the manufacture of the polymer and in the operation of the printer. PLA had contributions of the materials and processes similar to PETG, excluding the xylene production process, which does not exist in the PLA life cycle, including the use of mineral coal in the drying of corn grains for later production polymer. Lamnatou et al. (2019) showed significant contributions to the PETG manufacturing process in the category of non-renewable energy when using the ReCiPe method to study environmental impacts in a photovoltaic solar panel.

Regarding the carcinogen impact category, the substances with the greatest contributions to both PETG and PLA life cycles were aromatic hydrocarbons. Emissions from these hydrocarbons were mainly related to the production of xylene for the PETG and the production and burning of natural gas to supply energy in the PLA's extrusion processes. Cerdas et al. (2017) showed a significant impact on PLA extrusion and molding processes in relation to the emission of carcinogens.

In concern to the respiratory inorganic impact category, the emission of particulate material <2.5 μm (PM2.5) was the major contributor in both analyzed life cycle. The production processes of filaments and the use of electricity for printing the parts contributed significantly to both life cycles by the emission of MP2.5. However, the existence of the grain drying process in the PLA life cycle resulted in worse environmental performance in this category for it. Finally, the global warming impact category was mainly related to the emission of CO_2 in both life cycles. These emissions were related to the production processes of the filaments and the consumption of electricity in the use of the printer. Besides, the greenhouse gases contributed to the significant environmental impact in the production of xylene (PETG life cycle) and corn production (PLA life cycle).

In summary, it was realized that the impact categories of global warming, respiratory inorganic, and non-renewable energy were mainly related to the production process of electricity and thermal energy (burning), comprising the manufacturing phase of the filaments. For these three categories, it was found that the PLA life cycle obtained greater environmental impacts in the impact categories respiratory inorganics and global warming, while for the PETG life cycle, the impact category with the greatest environmental impact was the non-renewable energy.

Although these three categories have similarities in relation to their sources of potential environmental impact, there was a reversal of position between PLA and PETG. This should be due to the presence of fossil fuels in the production of xylene for the PETG life cycle and the existence of agriculture for the PLA life cycle. Cerdas et al. (2017) evidenced this high impact of agriculture activity in relation to global warming for PLA.

For the financial performance based on MFCA, the results of the comparative evaluation for the costs of products and waste are presented in Fig. 3.27.

We can see in Fig. 3.27 that the PETG is presented a greater cost of product than PLA. Thus, regarding the results of Sect. 3.8, it can be said that the lower loss of mass and also the higher unit cost of PETG acquisition made it present this better

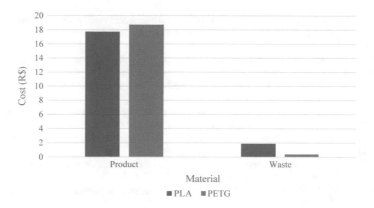

Fig. 3.27 Total cost of product and waste for each filament. *Source* The authors (2020)

performance. Consequently, the greater loss of material in the PLA printing process caused it a higher cost of waste, obtaining a worse financial performance. Despite this, it was realized that the biggest contribution to this performance came from the cost of the system (labor). That is, the workers generate waste instead of the product. This highlight of labor was due to the low cost of materials and energy involved in 3D printing when compared to the labor cost.

In summary, we can say that PETG had a worse environmental performance based on the LCA, and the PLA had a worse financial performance based on the MFCA. Analyzing better these results, PETG had a total environmental impact of 17.8% higher than PLA based on LCA (17.9 µPt of the PLA and 21.1 µPt of the PETG). However, based on MFCA, PLA presented an inefficiency in the transformation of raw material into a product (higher cost of waste) 733% higher than PETG (R$0.235 of the cost of waste for the PETG and R$1.723 of the cost of waste for the PLA). Therefore, we consider the PETG obtained better financial-environmental performance than PLA, given the greater performance discrepancy observed in the analysis by MFCA.

References

ACV BRASIL, *SIMAPRO* (2020). Available: https://www.acvbrasil.com.br/software/simapro. Accessed in: 12 Jan 2020

E.L. Cano, J.M. Moguerza, A. Redchuk, Process mapping with R, in *Six Sigma with R*. (Springer, New York, NY, 2012), pp. 51–61

A. Carvalho et al., From a literature review to a framework for environmental process impact assessment index. J. Clean. Prod. **64**, 36–62 (2014)

F. Cerdas et al., Life cycle assessment of 3D printed products in a distributed manufacturing system. J. Ind. Ecol. **21**(S1), S80–S93 (2017)

A.N. Dibofori-Orji, S.A. Braide, Emission of NOx, Sox and CO from the combustion of vehicle tyres in an abattoir. J. Nat. Sci. Res. **3**(8), 60–62 (2013)

N.K. Dixit, R. Srivastava, R. Narain, Comparison of two different rapid prototyping system based on dimensional performance using grey relational grade method. Procedia Technol. **25**, 908–915 (2016)

I. Gibson, D. Rosen, B. Stucker, *Additive Manufacturing Technologies: 3D Printing, Rapid Prototyping, and Direct Digital Manufacturing* (Springer, Berlin, 2014)

M. Goedkoop, R. Spriensma, *The Eco-indicator 99: A Damage Oriented Method for Life Cycle Assessment, Methodology Report* (PRé Consultants, 2000)

I.L.C.D. Handbook, Analysis of existing environmental impact assessment methodologies for use in life cycle assessment, in *Joint Research Center-European Commission* (2010)

ISO 14044, *Environmental Management—Life Cycle Assessment—Requirements and Guidelines* (International Organization of Standardization, 2006)

ISO 14051, *Environmental Management—Material Flow Cost Accounting—General Framework* (International Standard 14051, Geneve, 2013)

O. Jolliet et al., IMPACT 2002+: a new life cycle impact assessment methodology. Int. J. Life Cycle Assess. **8**(6), 324 (2003)

R.M. Jou et al., Biogenic fraction in the synthesis of polyethylene terephthalate. Int. J. Mass Spectrom. **388**, 65–68 (2015)

G.P. Kumar, S.P. Regalla, Optimization of support material and build time in fused deposition modeling (FDM), in *Applied Mechanics and Materials*, (Trans Tech Publications, 2012), pp. 2245–2251

C. Lamnatou, M. Smyth, D. Chemisana, Building-integrated photovoltaic/thermal (BIPVT): LCA of a façade-integrated prototype and issues about human health, ecosystems, resources. Sci. Total Environ. **660**, 1576–1592 (2019)

S. Madival et al., Assessment of the environmental profile of PLA, PET and PS clamshell containers using LCA methodology. J. Clean. Prod. **17**(13), 1183–1194 (2009)

O.A. Mohamed, S.H. Masood, J.L. Bhowmik, Mathematical modeling and FDM process parameters optimization using response surface methodology based on Q-optimal design. Appl. Math. Model. **40**(23–24), 10052–10073 (2016)

ÔMICROM, *Calibrador de Folga* (2018). Available: http://omicrom.com.br/geral/calibrador-de-folga/. Accessed: 12 Jan 2019

A.C.B. Passuello et al., Aplicação da avaliação do ciclo de vida na análise de impactos ambientais de materiais de construção inovadores: estudo de caso da pegada de carbono de clínqueres alternativos. Ambiente Construído: Revista Da Associação Nacional De Tecnologia Do Ambiente Construído **14**(4), 7–20 (2014)

K.N. Shah, N.S. Varandani, M. Panchani, Life cycle assessment of household water tanks—a study of LLDPE, mild steel and RCC tanks. J. Environ. Prot. **7**(5), 760 (2016)

G.A. Silva, L.A. Kulay, in *Avaliação do ciclo de vida*, ed. by A. Vilela Junior, J. Demajorovic, *Modelos de ferramentas de gestão ambiental*: desafios e perspectivas para as organizações (Senac, São Paulo, 2006)

J.V. Spadaro, A. Rabl, Estimates of real damage from air pollution: site dependence and simple impact indices for LCA. Int. J. Life Cycle Assess. **4**(4), 229 (1999)

E.T.H. Vink, et al., Applications of life cycle assessment to NatureWorks™ polylactide (PLA) production. Polym. Degrad. Stability **80**(3), 403–419 (2003)

World Health Organization, *Air Quality Guidelines for Particulate Matter, Ozone, Nitrogen Dioxide and Sulfur Dioxide* (2005)

J.W. Zhang, A.H. Peng, Process-parameter optimization for fused deposition modeling based on Taguchi method, in *Advanced Materials Research*, (Trans Tech Publications, 2012), pp. 444–447

Chapter 4
Perspectives and Future Applications

3D printing is a technology that has come to revolutionize the way we relate to products. In addition, it also came to streamline production processes in organizations. With the growth of this technology, a great diversity of equipment, materials, components, software, and other inputs related to 3D printing has started to mirror the market in a significant way. Also, the possibilities of types of users of 3D printing had a considerable increase in variety, given that the technology can be used both for large organizations, which aim to use 3D printing with high efficiency, accuracy, and speed, and for residential users, which has a compact equipment for 3D printing of products in a more simplistic and even artisanal way.

This great diversity of possibilities encourages us to think about what would be the most financially and environmentally advantageous choices for each type of use? Despite the complexity of the issue, this book seeks to provide support for the technology user to carry out a financial and environmental analysis in order to choose the alternative that best suits him.

The decision support process takes place through the sequential use of two techniques, namely LCA and MFCA. The application of the techniques occurs through the step-by-step presented in Chap. 2 of this book, starting with the steps that are common to both methodologies, later being divided into activities in parallel and finally converging in the comparison between their results.

In this book, the option of using the methodology was to evaluate the financial-environmental performance of two different types of filaments, PLA and PETG, used in 3D printing by FDM technology. The results of the study made it possible to show that PETG presented a financial performance in the transformation of material and energy flows significantly higher than that of PLA. And even if PETG has a worse environmental performance by LCA, its economic performance by MFCA was proportionally better, being considered for this study the material with the best financial-environmental performance.

We emphasize that the analyzes carried out using the LCA and MFCA methodologies were individualized, and in this study, the choice of the filament with the

© Springer Nature Switzerland AG 2021

T. Y. Kamiya et al., *Environmental and Financial Performance Evaluation in 3D Printing Using MFCA and LCA*, SpringerBriefs in Applied Sciences and Technology, https://doi.org/10.1007/978-3-030-69695-5_4

best financial-environmental performance was made by comparing the difference in results between PLA and PETG for the two methodologies. In spite of this, as already mentioned in Chap. 3, these methodologies seek to support the users' decision in relation to a given choice, being in this study the choice of the filament for use in 3D printing.

Based on what was portrayed in this book, the authors suggest some analysis options, which can serve to enrich the content related to financial-environmental performance with the use of the 3D printer.

When it comes to additive manufacturing, many technologies are being used and many are under development. The most popular is FDM, as discussed in this book. For future studies, other technologies can be used, which use metal and resin as raw material, and perform a comparative analysis between the materials used in the same technology and using different raw materials, as well as perform the comparison between the technologies, expanding knowledge about economic-environmental assessment.

In this book, we selected PLA and PETG material due to the characteristics of the printer used. In future studies, other materials can be studied, such as ABS, polyester, thermoplastic polyurethane (TPU), as well as composites, in which mixtures of materials are worked to improve mechanical characteristics, for example. When printing the piece, it was printed only, but due to the size of the printing table, more than one piece can be printed in the same print, which makes this experiment interesting to determine if there is variation in the economic-environmental performance, when impressions vary in quantity.

Another important feature addressed in the book concerns the quality of the pieces, and in this book, we address the superficial quality. From the design of the piece, it is possible to obtain the level of complexity for its printing. Thus, parts that have sharp curves, or need media for printing impact the final result of the part, and as a consequence can have an impact on economic-environmental performance, due to post-processing required to maintain the desired quality.

Based on the use of LCA and MFCA methodologies, we created an evaluation sequence for this study. Other methodologies of economic-environmental analysis can be used, as well as different printing technologies, each one with its specificity, but maintaining the logical sequence of the activities.

In our scenario, the landfill was used as the destination of the piece. This approach can be expanded by conducting studies with the part recycling as a destination, considering realities relevant to the business market and the residential user. In this way, cross analyses could be carried out with the landfill and recycling in the destination scenario and the business and residential user market as variables in reality.

The MFCA methodology was used considering three cost centers. For a more detailed view, depending on the analysis to be carried out, more cost centers can be considered, such as post-processing that depends on external resources that can significantly impact the final result, in addition to applying other cost accounting methodologies environmental, such as the product life cycle cost assessment methodology.

It is interesting to seek greater regionalization of the LCI used for the filaments, in order to provide greater proximity to the 3D printing scenario in the country where the study is carried out. As for the characterization method, IMPACT 2002+ was used because it is the most used method today to assess environmental impact in multi-indicators for products, but it can be used with other methods, such as ReCiPe, CML, and Eco-indicator for comparisons and also seek greater proximity to the scenario of the research country.

Printed in the United States
by Baker & Taylor Publisher Services